Occasional Publication Number 24

Northern Environmental Disturbances

Peter Kershaw
Editor

A publication of the
Boreal Institute for Northern Studies
1988

Copyright © 1988 Boreal Institute for Northern Studies

All rights reserved

ISSN 00680303
ISBN 0-919058-69-8

Published by:
Boreal Institute for Northern Studies
The University of Alberta
Edmonton, Alberta
Canada T6G 2E9
Telephone (403) 492-2919
Electronic mail: BINS@UALTAMTS
FAX: (403) 492-1153
Telex: 037-2979

Cover photo: Overview of the experimental simulation of a transport corridor (SEEDS) amid seismic lines near Fort Norman, Northwest Territories.
(Photo by Bonnie Gallinger)

Northern Environmental Disturbances

Peter Kershaw, Editor

Table of Contents

Introduction	1
Government's Role Peter Kershaw	3
Environmental Impact Assessment and Review in the North P.J.B. Duffy	5
Recent Industrial Experience with Mitigation and Reclamation of Northern Disturbances Peter Kershaw	9
Reclamation of the Norman Wells Pipeline Donald M. Wishart	11
Revegetation of Northern Disturbances Peter Kershaw	29
The Potential of Native Populations of Grasses in Northern Revegetation Manivalde Vaartnou	31
Revegetation in the High Arctic: Its Role in Reclamation of Surface Disturbance L.C. Bliss and N.E. Grulke	43
Ecosystematic Research Peter Kershaw	57
The Use of Controlled Surface Disturbances in the Testing of Reclamation Treatments in the Subarctic G.P. Kershaw	59

Introduction
G.P. Kershaw
Department of Geography
University of Alberta
Edmonton, Alberta T6G 2H4

This volume contains papers that were presented at a workshop during the Boreal Institute for Northern Studies twenty-fifth anniversary conference, *Knowing the North: Integrating Tradition, Technology and Science*. It was held in Edmonton between 20 and 22 November 1986. Within this workshop, entitled "Northern Environmental Disturbances: Mitigation and Reclamation," there were eight papers presented, five of which are contained in this volume.

Northern development has been a topic of discussion at a number of conferences. However, only recently have large-scale projects been initiated with mitigation and reclamation of disturbances as a focus of concern. The objective of this workshop was to draw together representatives from government, industry, and universities in an effort to review the current status of reclamation and mitigation of disturbances associated with the current development of Canada's North.

During the late 1960s and early 1970s an explosion in hydrocarbon and mineral exploration and prospecting activities occurred in northern Canada. At this time environmental issues were often at the forefront of public discussion concerning northern development. Governments were enacting legislation to reduce or mitigate environmental disruptions associated with developments and, in the North, the Territorial Land Use Regulations were finalized in 1972. This legislation along with others such as the Quartz Mining Act provide the legal basis for permit granting in the North. Hearings and subsequent agreements on the James Bay Project emphasized environmental impacts. During the mid-1970s the Mackenzie Pipeline inquiry and, later, hearings on the completion of the Dempster and Mackenzie Highways considered environmental impact mitigation. Much of our current understanding of the environment of the Mackenzie basin, and northwestern Canada in general, is founded on the work done in the 1970s in response to environmental concerns associated with these projects. More recently the expansion on the Norman Wells extraction facilities, the Norman Wells pipeline and the Beaufort Sea exploration operations have provided impetus for additional refinement of our environmental data base and the mitigation efforts associated with these developments.

Government's Role
Peter Kershaw
Editor

In most of Canada's North, environmental concerns remain within the federal jurisdiction. Little land is directly controlled by territorial and local government bodies. However, wildlife resources are within the administrative control of territorial governments. In addition to this split in government jurisdiction over environmental concerns there are numerous interest groups that have traditionally been active in the decision-making process as it affects proposed northern developments. These groups include the Canadian Arctic Resources Committee, the Committee for Original Peoples Entitlement, Dene Nation, Yukon Conservation Society, Chambers of Commerce, and many other agencies and organizations. At the local level one will find a number of interest groups that may include the Band Council, Hamlet Council, Meteis Association, Hunters' and Trappers' Association, Development Corporations, and Chamber of Commerce to name the most common ones. Indeed, one can easily be lost in the number and diversity of groups expressing opinion and expert testimony concerning proposed or operating development projects.

To assume that each of these interest groups and governing bodies have only one point of view is an oversimplification of reality. It is often possible to have differences of view points as to the merits and problems associated with specific mitigation and reclamation procedures. Furthermore, the agencies involved may have subgroups with different and potentially conflicting mandates. This results in extremely complex cases related to development proposals and specific mitigation and reclamation practices associated with major and minor projects alike. Pat Duffy's paper addresses some of these concerns and presents discussion on the specific projects that have undergone the Federal Environmental Assessment Review process.

Environmental Impact Assessment and Review in the North

P. J. B. Duffy, Director,[*]
Northern Region,
Federal Environmental Assessment Review Office,
Ottawa, K1A 0H3

Northern development in the last twenty years has been characterized by major transportation facilities, oil and gas exploration, major and middle-sized mining development with the attendant immigration of migrant workers and establishment of services. Examples of projects that have reached the development stage include the Aishihik Hydro Project, the Dempster and Mackenzie Highways, Beaufort hydrocarbon exploration, the Norman Wells Oilfield development and pipeline project, Panarctic hydrocarbon exploration in the High Arctic, and the Nanisivik and Strathcona Mines (Figure 1). A number of proposed projects have undergone extensive planning; examples are the Mackenzie Valley gas lines, Alaska Highway Gas and Dempster pipelines, Polargas Gas pipelines, Lancaster Sound hydrocarbon exploration, Arctic Pipeline Project, Mackenzie River Dredging, Great Bear River Hydro, and the Mackenzie Delta gas processing facility (Figure 1).

In the absence of organized and accessible data bases, large projects have required accelerated baseline information gathering and analysis in the planning stages. The scientific community and its sponsors have invested large resources of time, commitment, and funds to bring adequate information forward in a timely way for planning and decision-making. Engineering and economic analyses have proceeded with the added dimension of environmental impact assessment to identify potential project effects on the environment and suggest measures to reduce or eliminate such disturbances.

Since 1974 numerous northern projects of all sizes have undergone environmental impact assessment and review under the Federal Environmental Assessment and Review Process. All federally sponsored projects and development activities are now assessed in this way, as have been certain private sector projects, such as the Alaska Highway Gas Pipeline, Lancaster Sound Offshore Drilling and Arctic Pilot Projects. The largest proportion of these have been given approval by specialist committees charged with such responsibility after preliminary environmental screening. A smaller proportion required further investigation of potential effects and suitable mitigation measures. The resulting reports (initial environmental evaluations) contained recommendations prepared by the federal department responsible for the regulation of activities in a particular sector of the environment, with the advice of other government departments.

To date, a very small proportion of the projects have had potential for significant environmental effects requiring public review by a panel of specialists. Such a review normally involves community meetings in the

---

[*]Current address: 5839 Eagle Island, West Vancouver, B.C. V7W 1V6.

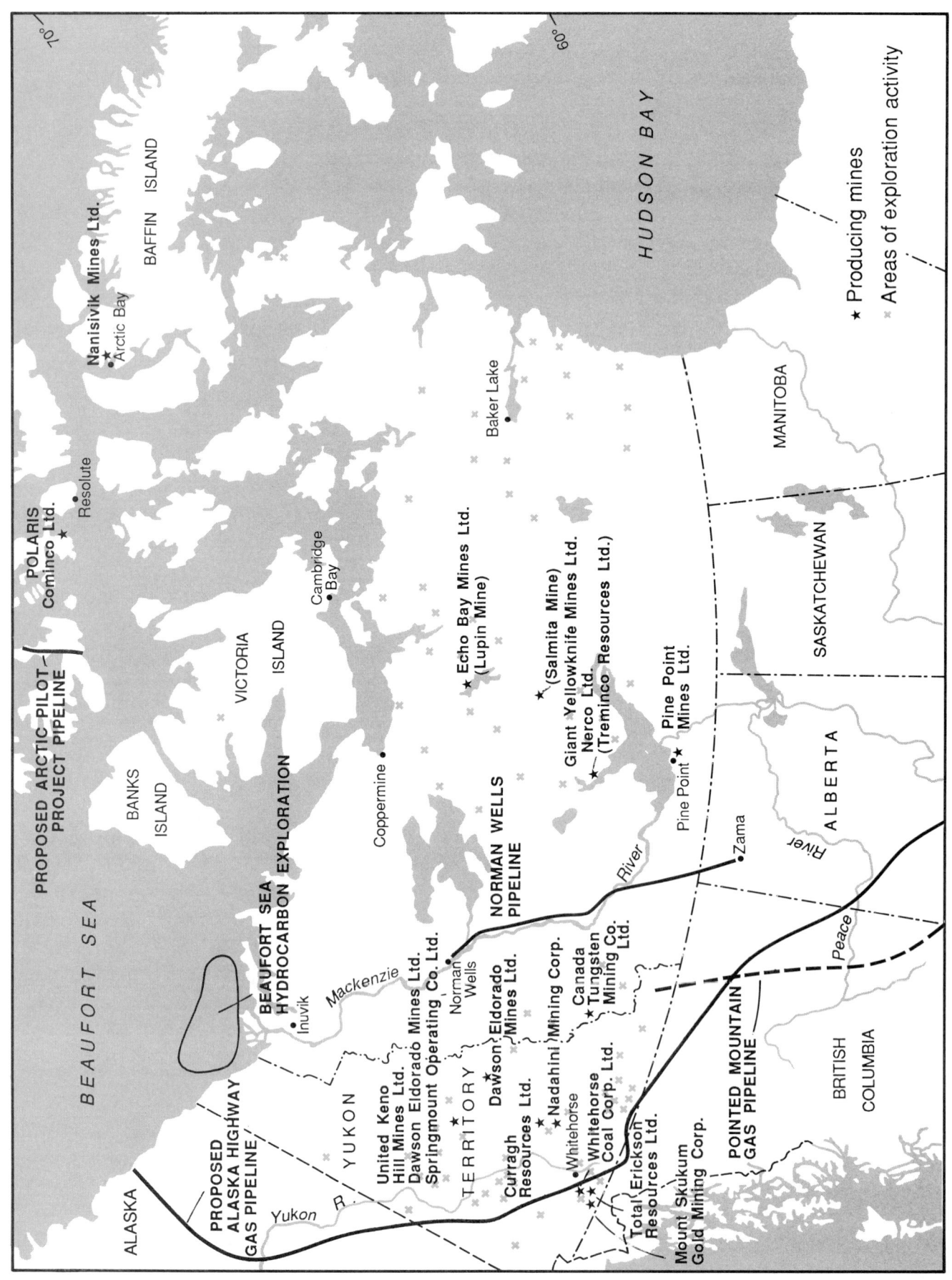

project area and technical hearings at a major center which permit a detailed examination of problems and issues by project developers, government reviewers, public groups, and interested individuals.

The environmental assessment of potentially significant effects focusses on the problem areas requiring priority attention. This occurs at all stages; from screening, through the initial environmental evaluation, to the public review panel stage. This environmental assessment process provides an integrative mechanism for the scientific disciplines and review participants to examine the issues systematically, placing emphasis on the critical ones, and in some cases, solving problems in the process. Examples can be found in pipeline routing — effects on wildlife habitat (Alaska Highway Gas Pipeline), the effects of artificial island construction on Mackenzie River fish species and habitat (Norman Wells Oilfield Development and Pipeline), and noise effects on marine mammals (Lancaster Sound Drilling).

A Weak Point in Project-Specific Assessment

The comparative absence of regional land and resource plans in the North, and long lead times required to develop solutions to some scientific problems are weak points in the project-specific approach to effects on the environment. That is, a predicament arises in cases where new technology needs to be applied to unique northern situations, such as pipelines in permafrost. The process of problem analysis and solution, including pilot project stages and ultimate approvals of government agencies such as the National Energy Board, may jeopardize the economics of proposed large-scale projects because the scientific investigations have insufficient time to deliver reliable results for project approaches and commissioning.

An example can be found in the case of large-diameter cold-gas pipelines considered for the Yukon and lower Mackenzie Valley. Potential for frost heave in transitional permafrost zones make reliable pipeline engineering design and construction a complex and expensive exercise. A design solution was developed by Foothills Pipelines (Yukon) Limited in 1982, for use in the Yukon section of the Alaska Highway Gas Pipelines. However, this was five years after the initial application by the company.

This predicament demonstrates the clear need for an ongoing survey and research program to focus on the unresolved and critical scientific problems which may impede planned developments in the North. In this regard, the present initiative on northern land use planning may help to identify and describe source problem areas. The Land Use Planning Commission for the Northwest Territories will conduct community and technical hearings, which will aid in identifying priority research requirements. As well, the programs of the Environmental Studies Revolving Fund, the Boreal Institute, the Science Institute of the Northwest Territories, the Yukon Science Institute, Northern Oil and Gas Action Program (NOGAP), and of other private, government, and industrial institutions will yield useful information on this predicament. However, the problem will remain until lead times for research are long enough to permit solutions to be tested in time for environmental reviews and subsequent approvals.

References

Berger, T.R. 1977. Northern Frontier Northern Homeland: The Report of the Mackenzie Valley Pipeline Inquiry. Two vols. — Supply and Services Canada, Ottawa.

Duffy, P.J.B. (ed.). 1986. Initial Assessment Guide: Federal Environmental Assessment and Review Process. — Federal Environmental Assessment Review Office (FEARO), Ottawa.

FEARO. 1977. Alaska Highway Pipeline: Report of the Environmental Assessment Panel. — Federal Environmental Assessment Review Office, Hull, Quebec.

— 1978. Shakwak Highway Project: Report of the Environmental Assessment Panel. — Federal Environmental Assessment Review Office, Hull, Quebec.

— 1978. Report of the Environmental Assessment Panel: Eastern Arctic Offshore Drilling, South Davis Strait Project. — Federal Environmental Assessment Review Office, Hull, Quebec.

— 1979. Report of the Environmental Assessment Panel: Lancaster Sound Drilling. — Federal Environmental Assessment Review Office, Hull, Quebec.

— 1979. Alaska Highway Gas Pipeline, Yukon Hearing: Report of the Environmental Assessment Panel. — Federal Environmental Assessment Review Office, Hull, Quebec.

— 1980. Arctic Pilot Project (Northern Component): Report of the Environmental Assessment Panel. — Federal Environmental Assessment Review Office, Hull, Quebec.

— 1980. Lower Churchill Hydroelectric Project: Report of the Environmental Assessment Panel. — Federal Environmental Assessment Review Office, Hull, Quebec.

— 1981. Norman Wells Oilfield Development and Pipeline Project: Report of the Environmental Assessment Panel. — Federal Environmental Assessment Review Office, Hull, Quebec.

— 1981. Alaska Highway Gas Pipeline, Routing Alternatives Whitehorse/Ibex Region: Report of the Environmental Assessment Panel. — Federal Environmental Assessment Review Office, Hull, Quebec.

— 1982. Beaufort Sea Environmental Assessment Panel, Interim Report. — Federal Environmental Assessment Review Office, Hull, Quebec.

— 1982. Alaska Highway Gas Pipeline, Technical Hearings: Final Report of the Environmental Assessment Panel. — Federal Environmental Assessment Review Office, Hull, Quebec.

— 1984. Beaufort Sea Hydrocarbon Production and Transportation: Final Report of the Environmental Assessment Panel. — Federal Environmental Assessment Review Office, Hull, Quebec.

Recent Industrial Experience with Mitigation and Reclamation of Northern Disturbances
Peter Kershaw
Editor

Exploration and transport corridors have been the most extensive types of disturbances in the North. Seismic lines crisscross most of northwestern Canada. Telecommunication line, winter road, and pipeline rights-of-way, as well as highways and access roads have extended surface transport facilities to many remote areas in this region. Settlements, mining, and hydrocarbon drilling sites have been linked by these lines and there are numerous proposals for additional developments in Canada's North. Industrial-sector interest groups have instigated many of these ventures and more are being proposed. This is a natural evolution of Canada's North as it becomes more important as an area of resource exploitation. Accompanying these developments are the environmental disturbances that typify any land-based construction and operation activities. Significant northern phenomena such as permafrost and environmental extremes in weather, however, mean that development-related activities which would have little consequence in less extreme environments, can, in the North, cause severe and irrevocable changes ranging from short-term to long-lasting; minor to major; and negative to positive environmental alterations. Although the evaluation of these changes is difficult and open to interpretation, they undisputedly result from development-related activities in Canada's North.

For the first time in the history of northern development we can objectively evaluate the environmental responses to disturbances. The current technologies associated with monitoring of these changes have evolved to a state where accurate measurements can be achieved. Furthermore, the commitment by industry, government, and research scientists to the evaluation of these changes is real and is being acted on. Don Wishart's paper on the reclamation practices associated with one of the most recent large-scale northern development projects is an example of the current state of application of principle to practice.

Reclamation of the Norman Wells Pipeline

Donald M. Wishart
Interprovincial Pipe Line Limited
P.O. Box 398
Edmonton, Alberta T5J 2J9

Introduction

In the spring of 1985 Interprovincial Pipe Line (NW) Ltd. began operation of a pipeline that extends from Norman Wells, Northwest Territories to Zama, in northern Alberta (Figure 1). This was the first completely buried pipeline constructed north of the 60th parallel and, as such, is considered a pilot project to test various northern reclamation theories.

Reclamation is a critical element of any pipeline development and it is obvious that the future of resource development in the North will be tied to the possibility of mitigating environmental damage. Inadequately reclaimed sites can contribute to loss or reduction of biological productivity, degradation of water quality, and terrain instability that ultimately could threaten the integrity of the pipeline. In the northern pipeline context, reclamation becomes problematic due to the presence of potentially sensitive permafrost soils.

This paper addresses the rationale, evolution, and application of various reclamation measures used on the Norman Wells pipeline. While the development of slope stabilization designs included many of the reclamation procedures discussed below, it is beyond the scope of this paper to fully address slope design criteria.

Norman Wells Pipeline Description

The Norman Wells pipeline is a 323 mm diameter, crude oil pipeline that is 869 km in length. The pipeline is buried throughout its entire length at an average depth of 0.9 m. Right-of-way clearing began in the winter of 1983 and pipe was laid over the winters of 1984 and 1985. The right-of-way was cleared to a width of 25 m and, wherever possible, the pipeline was located in previously disturbed alignments such as seismic lines or abandoned telegraph lines. Pipeline operations began in April 1985. The oil is cooled to below 0°C before entering the pipeline and thereafter is allowed to assume the ambient thermal conditions of the ground.

For 526 km south of Norman Wells, the pipeline parallels the Mackenzie River. Along this portion of the route, which has numerous valleys and high flat terraces, the pipeline crosses approximately 160 identifiable drainage courses. South of the Mackenzie River, the route is generally flat and poorly drained.

Although permafrost occurs over a large proportion of the pipeline route (Nixon *et al.* 1983), it varies in distribution from approximately 85% of terrain for the first 150 km. south of Norman Wells to less than 20% in the southern portion of the pipeline. A 1 m active layer overlies permafrost 50 m thick

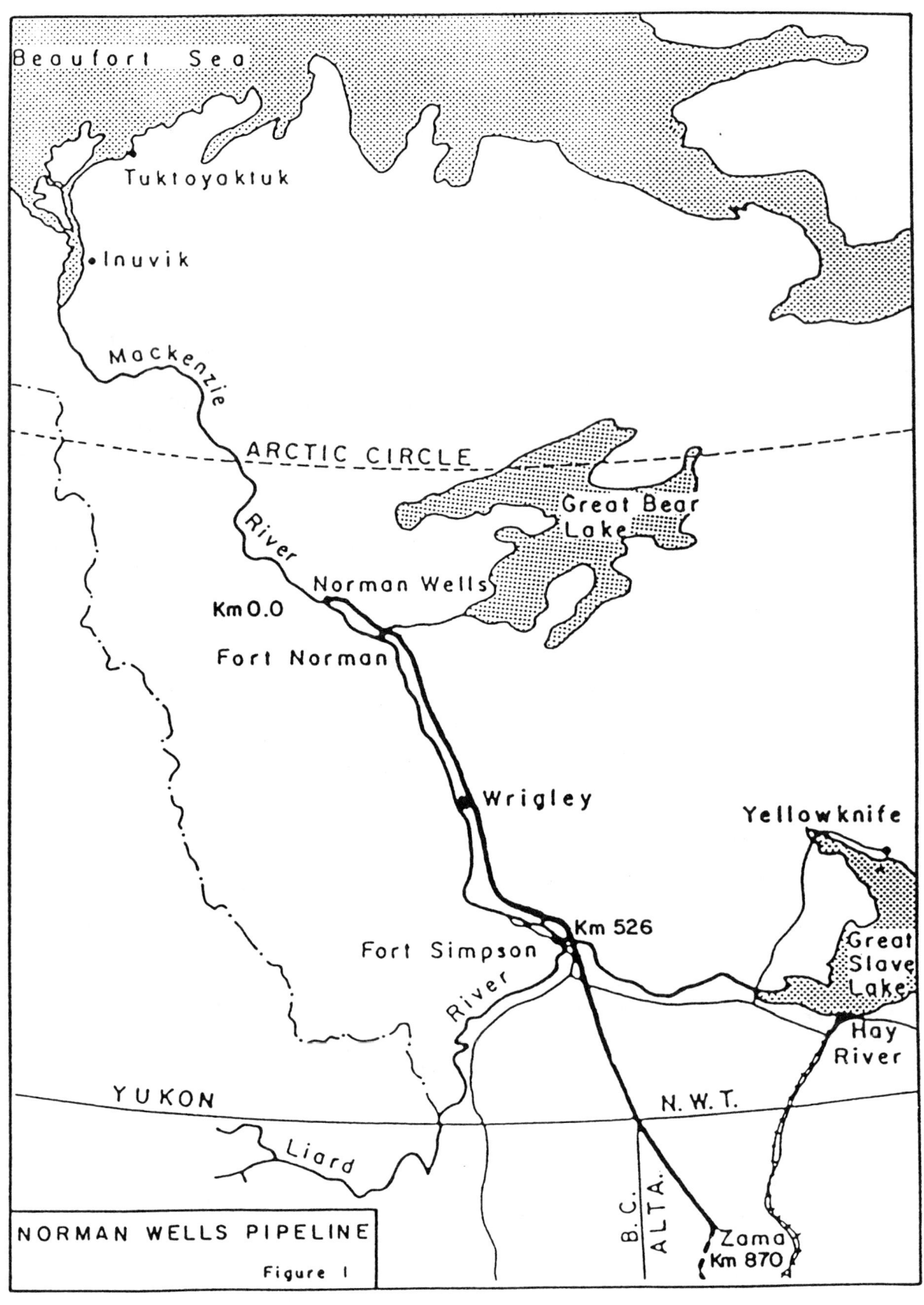

NORMAN WELLS PIPELINE

Figure 1

near Norman Wells. The active layer increases to 3 m. near Zama where permafrost occurs in scattered patches of 1 m in thickness.

The vegetation along the route is dominated by open coniferous forests of dwarfed *Picea mariana* (Mill.) BSP and *Larix laricina* (Du Roi) K. Koch with intermittent areas of bogs and fens. Where topography is rolling, well-drained sites support *Picea glauca* (Moench) Voss, *Populus tremuloides* Michx. and *Pinus banksiana* Lamb.

Surface Drainage Control

An important — and to date the most difficult — element in the reclamation process for the Norman Wells pipeline has been the control of surface drainage (Wishart and Fooks 1985). During construction it was necessary to grade the surface of the right-of-way to create a reasonably level driving surface that accommodated up to 250,000 man-km travel each day. While the extent of grading was minimized in order to preserve the organic mat, much of the resulting surface was left in a denuded, uniform condition which provided little resistance to water velocity.

Differential settlement, in response to heterogeneity of underlying geological materials, will eventually reproduce a microhummocky terrain that is less sensitive to surface drainage problems, and a stable vegetation mat will ultimately provide the most secure means of controlling surface drainage. In the interim, however, it has been necessary to utilize various mechanical erosion control measures, ranging from simply breaching the backfill mound to construction of complex drainage systems.

Facilitating Cross-Drainage

Ditching operations associated with buried pipeline systems result in an expansion of soil volumes relative to consolidated predisturbance conditions. The expanded volumes are a result of an increase in size and number of macro-pore spaces in the disturbed regolith. In order to accommodate subsequent consolidation and settlement along the Norman Wells pipeline, the backfilled materials were mounded over the ditch line to heights of up to 2 m above mean ground level.

Wherever natural drainages crossed the right-of-way, a continuously mounded backfill would have presented a drainage barrier. In order to maintain cross right-of-way drainage, the backfill was leveled for short distances wherever cross drainage was evident. Where cross drainage was poorly defined, runoff was trained across the ditch line at periodic stable locations varying up to 500 m apart.

Two problems arose with the implementation of this design. First, the breaches were not always placed to take advantage of minor drainages which were difficult to identify. As a result, water occasionally was trained to flow along the backfill mound. Second, the breached area frequently subsided and water pooled on the ditch line rather than flowing across the right-of-way. In both instances, water tended to flow down the ditch line rather than across the right-of-way.

In the second year of mainline construction, these problems were addressed with changes in the design of the breaches and in procedures used in both the pre-construction and post-construction periods. In areas of poorly defined drainage, a preconstruction survey was undertaken during the thaw season to identify where best to locate the breach in the backfill mound. The locations were carefully flagged in the field. During pipeline construction, the breached area was backfilled with either sand bags or compacted granular-backfill to minimize future subsidence. A short diversion berm was also placed on the down-slope face of the backfill mound to prevent water movement down the ditch line. Where cross drainage was not identified, the backfill mound was not breached during construction. Instead, breaches were made by hand where required in the first thaw season after construction. Most cross-drainage problems were resolved by these design changes.

Water Diversion

The most effective means of controlling surface drainage is to divert the flow off the right-of-way to stable undisturbed terrain. To achieve this, drainage ditches were installed where mineral soils were graded below the original contour and the potential for interception of drainage was high. The drainage ditches were designed to ensure that the intercepted flow would be collected, taken down the right-of-way in a controlled, non-erosive manner and then directed off the right-of-way. The design specified rock armour from 100 to 300 mm in diameter to line the ditch and included the potential for a geotextile liner (Figure 2). To date, the drainage ditches have performed well.

Diversion berms were used extensively on the Norman Wells pipeline project to intercept and drain water off of the right-of-way. The diversion berms were constructed of sand-filled bags or native soil and geotextile fabric. Eventually, the bags will deteriorate and the surface of both sand bag berms and native soil berms will become stabilized by vegetation. The diversion berms were designed to have a 10° slope and the frequency of berm placement was a function of slope (Figure 3).

When backfilling, it was not possible to mound all of the excavated materials over the pipeline ditch. The angle of repose of some materials dictated that the base of backfill mounds exceeded 2 m in width relative to a 0.8-m-wide ditch. As the backfill materials consolidated in the thaw season following construction, the ditch line often subsided below the adjacent grade.

Berms installed during the first year of construction occasionally collapsed where they crossed the ditch line as a result of subsidence of the ditch. These collapsed diversion berms channelled all water collected above the ditch line over the collapsed portion of the berm and then directly down the ditch line. In order to alleviate the degree of settlement of the berm over the ditch line in the second year of construction, the berm was supported by using compacted granular backfill or sand bags in the ditch (Figure 4), and an additional layer of sand bags was placed over the ditch line to accommodate residual subsidence.

Watercourse Designs

Drainage control was carefully designed at all watercourses in recognition of their environmental sensitivity. At certain watercourses, all the techniques discussed above were utilized. Furthermore, in the watercourse, the ditch line was backfilled with granular material to the original grade of the stream bed. Rock armouring was placed on the stream banks (Figure 5).

Few problems were encountered on approach slopes. However, in the first thaw season after mainline construction there were repeated instances of ditch-line subsidence in the watercourse, on the banks, and on adjacent floodplains. The watercourse design was subsequently amended by increasing the height of the backfill mound on the watercourse banks and floodplains and expanding the use of armour. To date, ditch-line subsidence has been greatly reduced by implementing these measures.

Subsurface Drainage

The unconsolidated nature of backfill material and the surface of the pipe in the ditch

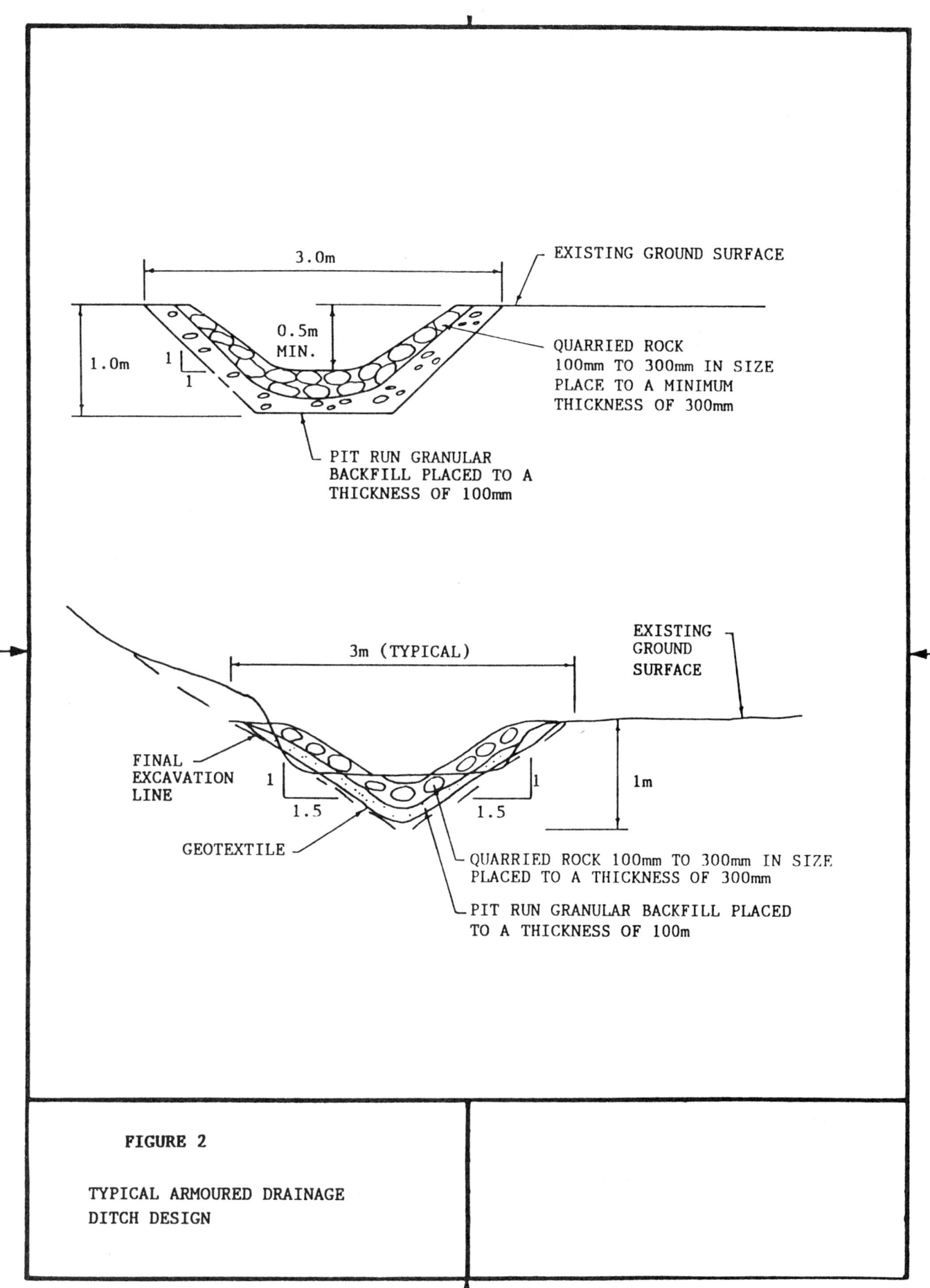

FIGURE 2

TYPICAL ARMOURED DRAINAGE DITCH DESIGN

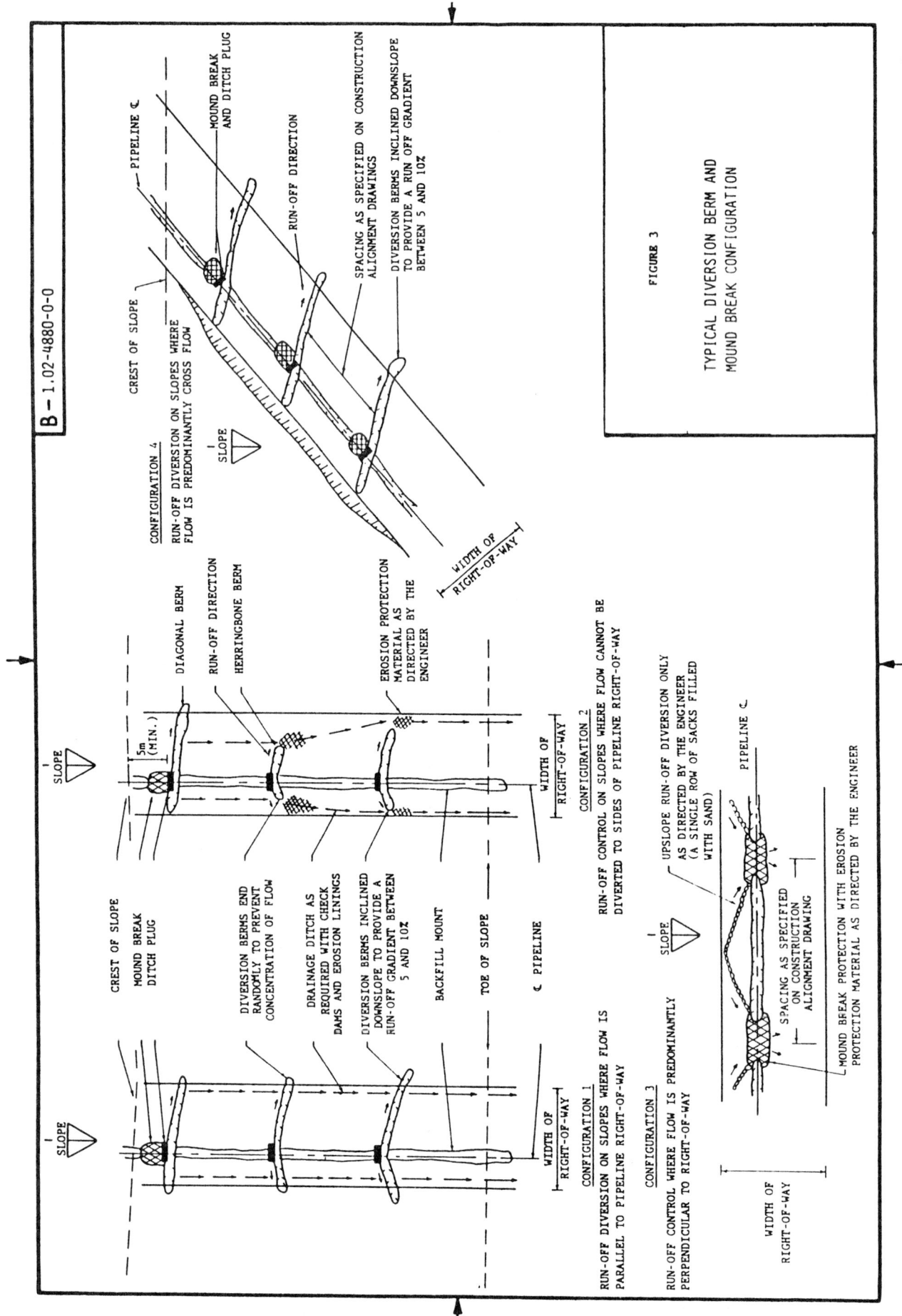

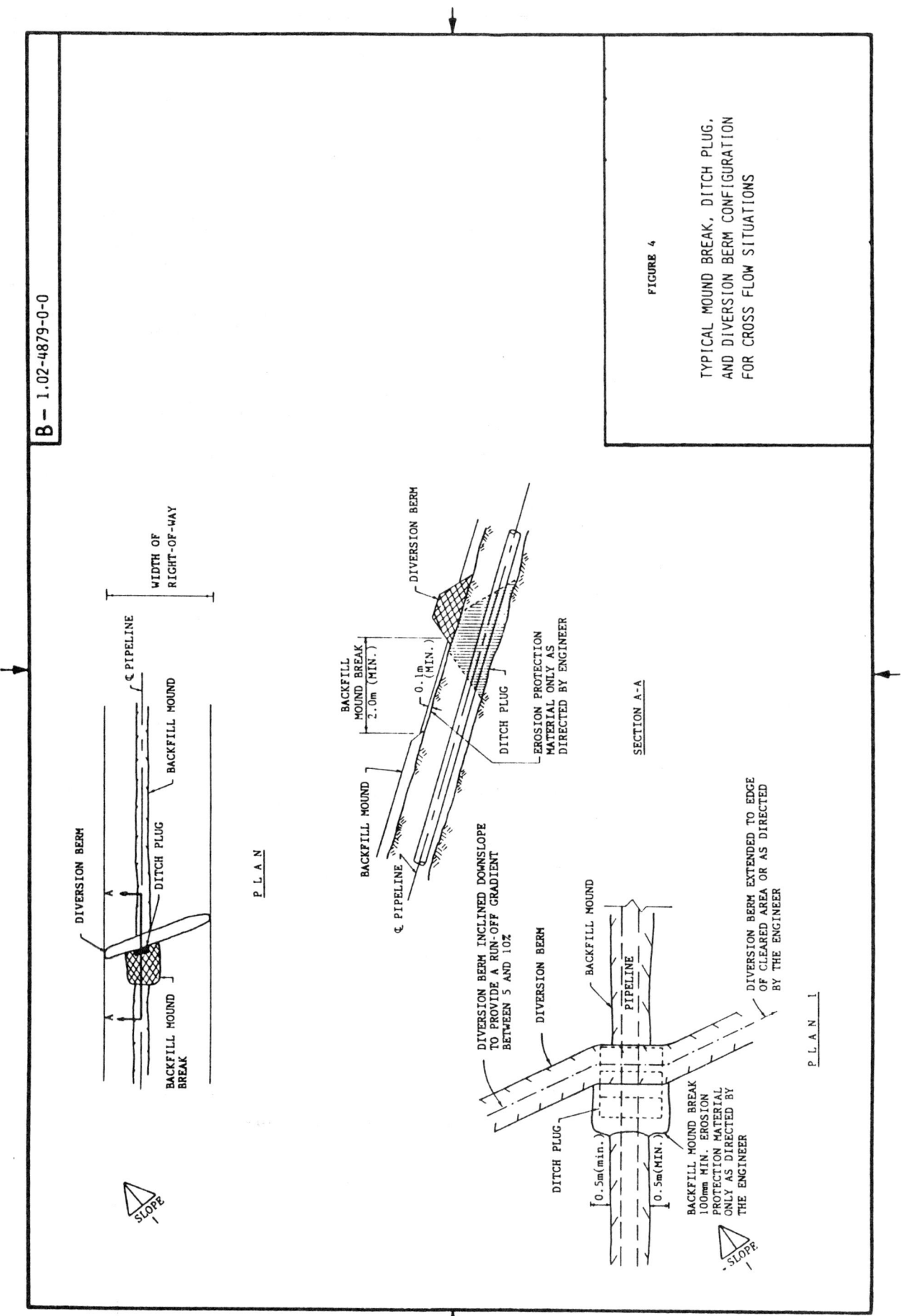

FIGURE 4

TYPICAL MOUND BREAK, DITCH PLUG, AND DIVERSION BERM CONFIGURATION FOR CROSS FLOW SITUATIONS

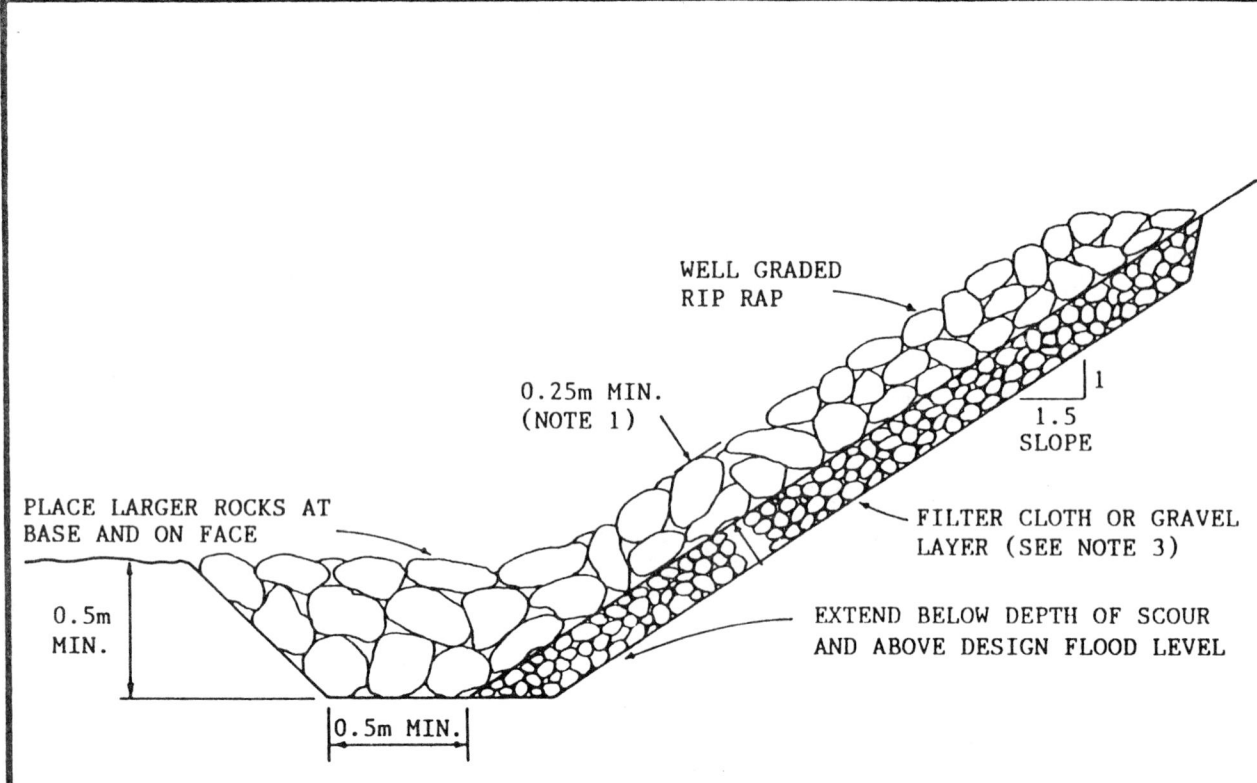

PROFILE

NOTES:

1. INSTALL RIP RAP TO A DEPTH APPROXIMATELY 1.5 TO 2 TIMES THE DIAMETER OF THE RIP RAP.

2. SIZE OF RIP RAP USED IS DEPENDENT UPON SLOPE OF BANK AND WATER VELOCITY. FLAT ROCKS ARE MORE EFFECTIVE THAN ROUND ROCKS BECAUSE THEY RESIST MOVEMENT BY WATER AND ICE.

3. INSTALL FILTER CLOTH OR GRAVEL LAYER IF WATER TURBULENCE COULD RESULT IN EROSION OF BANK MATERIAL BETWEEN LARGE ROCKS.

4. CONSTRUCT RIP RAP BOUNDARIES IN A MANNER SUCH THAT IT WILL NOT BE UNDERMINED FROM THE SIDE.

FIGURE 5

RIP RAP INSTALLATION FOR WATERCOURSE BANKS

offer a low friction seam which can facilitate high volume, high velocity, subsurface water flow. Where left uncontrolled, subsurface drainage can wash out the backfill material. The collection and concentration of water in the ditch could also lead to excess pore pressure and slumping, as could rapid thaw degradation in the case of ice-rich permafrost soils.

Ditch Plugs

In order to prevent water flow down the ditch line, ditch plugs were designed and installed to reduce velocity, to force subsurface flow up and out of the ditch where a surface berm directs the water off the right-of-way, and to physically restrain mass movement of soils down the ditch line. Ditch plugs are essentially a wall of low permeability that fills the ditch and surrounds the pipe. They were constructed of sand bags, bentonite, and geotextile (Figure 6) or polyurethane and bentonite. The plugs were keyed into the ditch wall and extended from the ditch floor to the original ground surface.

Initial difficulties arose where the plugs were not properly keyed to surface berms or where the surface berms subsided sufficiently to allow water to continue to flow down the ditch line. In addition, the plugs were not used efficiently on portions of the right-of-way with long gentle slopes. These problems were resolved by ensuring that all plugs were keyed at the surface into a diversion berm (Figure 4) and their use was extended on long gentle slopes.

Insulated Slopes

To maintain stability on ice-rich slopes, it was considered necessary to insulate the slopes to retard or prevent thaw (McRoberts *et al.* 1985). Wood chips were selected for the insulating medium largely because of their workability even in cold weather, and their ability to settle as thaw regression took place. Design thickness of insulation varied from 0.8 m to 1.8 m depending on degree of slope, soil composition, ice content depth of organic overburden, and width of right-of-way clearing.

While it is too soon to determine the long-term effectiveness of the wood-chip insulation, no stability problems have developed on ice-rich slopes and the wood-chip insulation appears to be functioning according to design predictions.

Revegetation

The removal of the tree canopy and subsequent increases in ground surface insolation will eventually thaw permafrost soils to a depth of about 12 m, unless they are mechanically insulated or chilled. The volume of soil occupied by water will decrease as water passes from solid to liquid phase and as the melting water drains from the soil. Over time, the right-of-way in areas of permafrost soils will subside relative to adjacent undisturbed terrain. As the ground surface subsides, the right-of-way will tend to channel water and the diversion berms will become increasingly less effective in directing water off of the right-of-way. On the Norman Wells pipeline, the long-term solution to this problem has been to establish a stable, erosion-resistant, vegetation mat which physically reduces surface water velocities and transpires large volumes of subsurface water out of the soil column. The vegetation can also enhance wildlife habitat.

In late winter immediately after construction, a broad-base seed mix capable of occupying a variety of micro-habitats was broadcast at a rate of 30 kg/ha along the right-of way and at associated construction sites. Seed was applied at 50 kg/ha on slopes. Large, level peat bogs were allowed to revegetate naturally because of their low potential for erosion. A nitrogen-phosphorus-potassium fertilizer blend (17-25-15) was

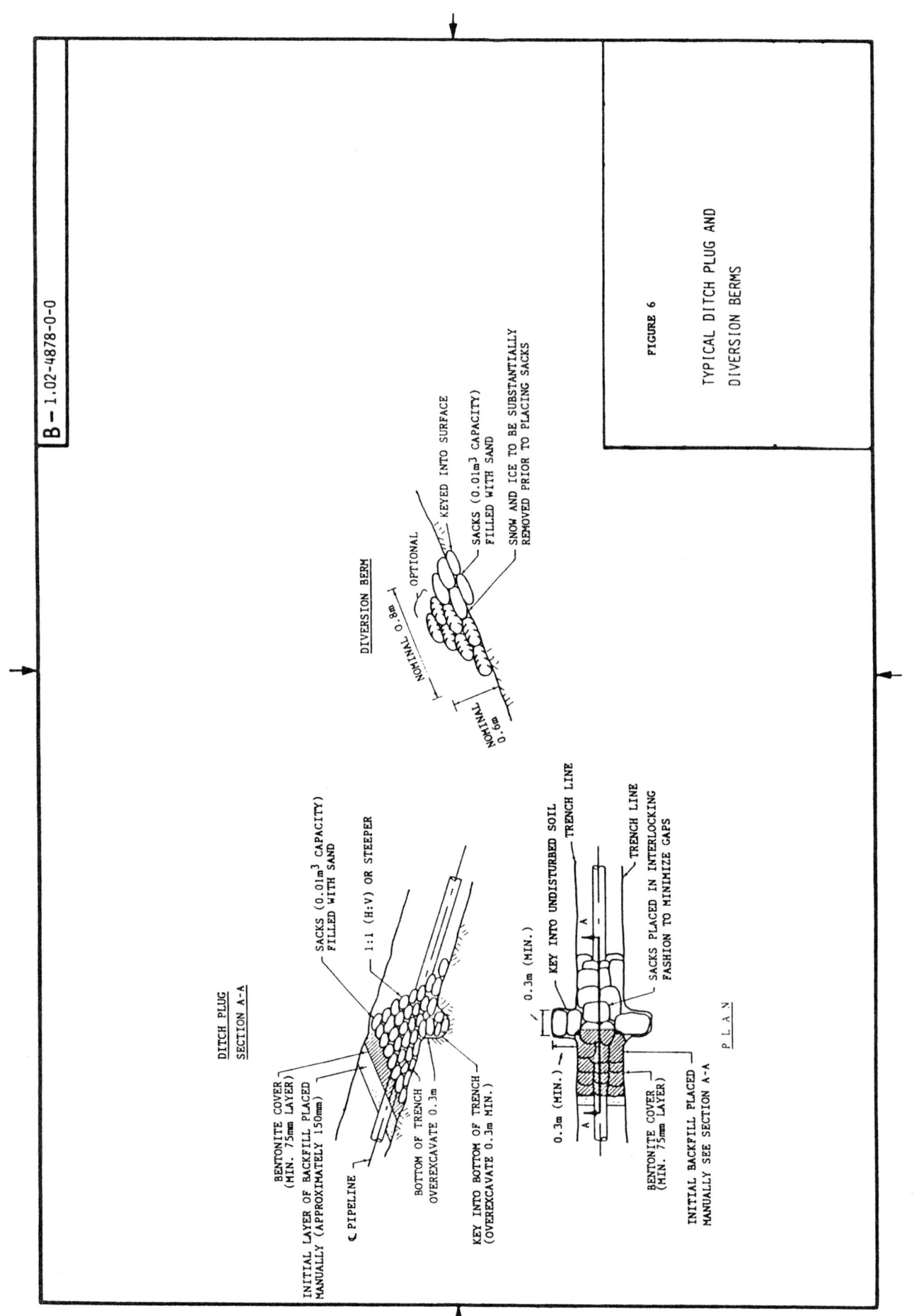

FIGURE 6

TYPICAL DITCH PLUG AND DIVERSION BERMS

Plate 1: Third growing season vegetation along Norman Wells pipeline

Plate 2: Third growing season vegetation on a heavily graded site

broadcast at 250 kg/ha wherever seed was applied. Since construction, seed and fertilizer have been applied at similar rates wherever revegetation was considered unsatisfactory. In total, approximately 45 tonnes of seed and 445 tonnes of fertilizer have been utilized on the Norman Wells pipeline project.

The seed mix used in the various seeding programs depended on availability of stock and short-term success as determined by quantitative revegetation monitoring conducted each growing season. In total, ten species have been utilized in different combinations to make five seed mixtures. All seed mixes were comprised of common agronomic grass species and, to ensure high seed quality and minimize potential for weed infestations, all seed utilized was graded Canada Number 1 or higher.

Revegetation of the right-of-way has been favourable (photographic plates 1 and 2). In two growing seasons, total plant cover averaged 44% along the right-of-way with approximately two-thirds of the cover provided by seeded grasses and one-third by native vascular plants (Hardy BBT 1986). In August 1986, only about 15% of the right-of-way had less than 25% ground cover. These sites generally occurred on the peat bogs which were not seeded. While presently poorly vegetated, no erosion or other significant environmental concerns have developed on these peat areas.

In order of decreasing abundance, Boreal creeping red fescue (*Festuca rubra* L.), creeping foxtail (*Alopecurus arundinaceae* L.), Revenue slender wheatgrass (*Agropyron trachycaulum* [Link] Malte) and Climax timothy (*Phleum pratense* L.) were the most successful of seeded species. Reed canary grass (*Phalaris arundinacea* L.) was locally abundant on hydric sites.

A trial shrub planting program was undertaken in the spring of 1985 to test the effectiveness of shrub cuttings to improve terrain stability at slopes and watercourse banks. Approximately 45% of cuttings survived after two growing seasons. Most cuttings planted on watercourse banks were washed away in the spring freshet and at one site several of the cuttings were buried by a shallow slump. While insufficient time has passed to derive conclusions concerning the long-term effectiveness of shrub plantings, it is apparent that the shrubs will not significantly contribute to terrain stability for several years.

Monitoring Programs

In order to ensure that any environmental or geotechnical problems associated with the pipeline are identified and mitigated as soon as possible, several, long-term monitoring programs have been initiated (Table 1).

Low-level aerial line-patrols of the entire right-of-way are carried out at least once each week. Ground inspections are conducted wherever undesirable terrain conditions are identified but, in most cases, remedial work is completed during line patrol.

Comprehensive geotechnical monitoring programs have been established in order to determine thaw settlement, frost heave, and slope performance. These programs include monitoring undertaken in cooperation with Indian and Northern Affairs Canada, with assistance from Energy, Mines, and Resources Canada. Geotechnical monitoring of slopes takes the form of aerial surveys, ground inspections, and instrumentation readings. Collectively, more than ninety sites have been instrumented with thermistors, piezometers, settlement plates and/or strain gauges. Instruments are generally read on a monthly basis.

In the late summer of 1984, annual monitoring of the right-of-way and other associated construction sites was initiated in order to

Table 1. Schedule of Post-Construction Monitoring Programs.

Program	Initiate	Terminate	Frequency
Standard Operational Monitoring			
—Aerial surveillance	1984	Lifetime*	Weekly
—Ground inspection	1984	Lifetime	Annual
Geotechnical Monitoring			
—Thaw settlement	1984	Lifetime**	Annual
—Frost heave	1984	Lifetime**	Four/Year
—Slope performance	1984	Lifetime**	Eight/Year
Aquatic Monitoring			
—Mackenzie and Great Bear river bed elevations	1985	Lifetime**	Annual
—All watercourses	1984	Lifetime	Weekly
Raptor Monitoring	1983	1987	Annual
Wildlife Monitoring			
—Right-of-way monitoring	1984	1988	Weekly
—Transect monitoring	1984	1987	Bi-annual
Revegetation Monitoring			
—Aerial survey	1984	Lifetime**	Annual
—Baseline data collection	1984	1985	Annual
—Permanent sample transects	1984	1987	Annual
—Revegetation trail sites	1984	1987	Annual
Aerial photography and ground truthing	1986	1988	Biennual

* "Lifetime" refers to operational lifetime
** Program may be discontinued if history of readings indicate an equilibrium situation.

assess revegetation success. Indian and Northern Affairs Canada is also conducting revegetation monitoring and the two programs have been designed to complement one another.

The revegetation monitoring program consists of four components: aerial surveys, baseline data collections, monitoring of permanent sample transects, and monitoring of revegetation trial sites. The aerial surveys are conducted to provide a log of revegetation success along the right-of-way. Documentation is qualitative: areas where revegetation is poor are noted and special emphasis is placed on reconnaissance at slopes or highly erodible sites. Permanently marked transects are monitored to provide a quantitative assessment of revegetation over time and twenty revegetation trial sites were established to determine the effectiveness of seeding and fertilizing.

Annual site inspections are conducted on all permanent watercourses crossed by the pipeline. The inspections consist of examinations of the physical condition of the streambed, extent of scour, the stability and condition of stream banks, an assessment of potential for watercourse sedimentation, and an assessment of whether the right-of-way is contributing to the degradation of water quality or aquatic habitat.

A program of aerial photography also has been undertaken to provide a chronological record of terrain and landscape conditions before, during, and after construction. Aerial photographs of the right-of-way have been taken in 1980 and annually since 1983; the next photographs will be taken in 1988.

Remedial Reclamation Programs

Most of the reclamation measures discussed above were implemented during construction in order to avoid or minimize environmental problems. However, a program of remedial measures has also been undertaken to repair or arrest problems as they develop. Remedial measures are implemented on a site by site basis and the degree of response is controlled by the severity of the problem and the means of access available.

In the absence of an all-weather road in proximity to most of the right-of-way, remedial measures during the thaw season are largely limited to heliocopter-supported hand operations with locally available or heliocopter-transportable construction materials. Caches of sand bags have been located at approximately 2 km intervals along the right-of-way to provide readily accessible erosion control materials. In addition, six major caches of seed and fertilizer have been located at maintenance bases along the right-of-way. Major remedial work has been restricted to the winter period when access by heavy machinery and larger work crews is feasible. The winter programs are intended to resolve chronic drainage control difficulties.

Most remedial efforts have been directed towards replacement or relocation of reclamation structures installed during construction. For example, several streambanks have been rearmoured with rock rip-rap, and diversion berms and ditch breakers have been replaced or installed where required. However, two types of problems have developed that were impractical or impossible to address with designs appropriate for pipeline construction.

Erosion on Low-Angle Slopes

Diversion berms have worked well on steep or well-defined slopes where drainage direction is easy to determine. However, on sites where the slope angle is low, or where general drainage tends to follow the right-of-way, water directed off of the right-of-way by berms often flowed back further down slope. On long, gentle slopes, the velocity of flow is low but the large volume of water has resulted in erosion problems, particularly along the unconsolidated ditch line.

As discussed above, a stable vegetation mat will provide long-term erosion control on these sites. However, where water cannot be effectively diverted using berms or similar structures, interim remedial measures have been taken to reduce velocity of water to non-erosive levels. Several techniques have proven successful. Sand bags, straw bales, and native soils have been placed across the right-of-way to create a series of check dams. The velocity of water down the right-of-way decreases and erosion has been greatly reduced. In some cases, water has pooled upslope of these dams and eroded materials have tended to settle. Redistribution of residual slash from clearing operations, or placement of flax, polypropylene, or jute matting have also proven successful at reducing water velocity and holding the soil in place.

Subsidence of the Pipeline Ditch

Where ice-rich soils were encountered, the removal of the insulating organic overburden during ditching operations, and exposure of low albedo mineral soils has resulted in more rapid thawing of ditch-line soils relative to adjacent soils. The more rapid thawing along the ditch line has aggravated subsidence caused by an inability to place all of the excavated materials over the ditch as described above. The subsided pipeline trench has tended to intercept and channel cross right-of-way drainage. Once flow is directed down the ditch line, erosion increases the problem.

The phenomenon of subsidence of the pipeline ditch has been the most extensive problem addressed since construction. To date, over 200 km of right-of-way have displayed some degree of subsidence of the pipeline ditch below the adjacent grade and approximately 80 km have had subsidence in excess of 20 cm in depth. Several techniques have been used to address subsidence of the ditch line.

Throughout the thaw season, new diversion berms have been installed to force water out of the subsided ditch and off the right-of-way. On low-slope angles, a series of check dams have been installed in the ditch to reduce water velocity and settle eroded materials. The check dams were made from various materials including sand bags, straw bales, and local timbers.

While the diversion berms and check dams have worked effectively for controlling erosion, they do not address the problem of drainage down the pipeline ditch in the long term. Wherever subsidence of the pipeline ditch exceeded 20 cm, granular material was excavated from borrow sites, transported to the right-of-way, and used to reestablish the surface of the ditch line at or above the adjacent grade.

Where the subsidence of the ditch line was less pronounced, a remedial seeding program was undertaken. The objective was to establish an erosion-resistant grassed drainage-way similar to roadside ditches in southern environments. Seeding was undertaken after spring thaw in order to avoid the large volume of flow that could carry seed from the site, while providing a sufficient growing season for vegetation to establish.

Both backfilling with granular materials and reseeding along the ditch line have proven effective. However, it is probable that new areas will have to be addressed in each of the next few years as ditch-line soils continue to consolidate and until thawing of soils adjacent to the ditch line reaches equilibrium with thawing in the ditch.

Conclusion

Considerable experience has been gained in

the field of reclamation of pipeline rights-of-way in northern environments since the first construction season. The evolution of designs based on theory to those designs proven effective and practicable in various field situations reflects an on-going process of improvement in northern environmental management. On the Norman Wells pipeline, reclamation problems will diminish over time as disturbed soils consolidate and the vegetation cover along the right-of-way becomes better established. Nevertheless, monitoring programs will continue throughout the operational life of the pipeline to ensure rapid implementation of remedial measures.

References

Hardy BBt Limited. 1986. 1986 revegetation monitoring of the Interprovincial Pipe Line (NW) Ltd. Norman Wells to Zama pipeline. Prepared for Interprovincial Pipe Line (NW) Ltd., Edmonton, Alberta.

McRoberts, E.C., Nixon, J.F., Hanna, A.J., and Pick, A.R. 1985. Geothermal considerations of wood chips used as permafrost slope insulation. Proceedings of Fourth International Symposium on Ground Freezing. Sapporo, Japan, pp. 305-312.

Nixon, J.F., Stuchly, J., and Pick, A.R. 1983. Design of Norman Wells pipeline for frost heave and thaw settlement. Third International Symposium on Offshore Mechanics and Arctic Engineering. ASME Trans. of J. of Ener. Res. Tech. Pap. No. 83-OMA-303.

Wishart, D.M. and Fooks, C.E. 1985. Norman Wells pipeline project, right-of-way drainage control — problems and solutions. Proceedings of the Northern Hydrocarbon Development Environmental Problem Solving Conference, Banff, Alberta, Sept. 24 - 26, 1985, pp. 209-218.

Revegetation of Northern Disturbances
Peter Kershaw
Editor

The necessity of re-establishing plant cover on northern disturbances has been accepted by project proponents, regulatory agencies, and concerned interest groups. In the North a surface plant cover moderates microclimates and affects soil climate. Vegetation reduces soil erosion by reducing ice-rich permafrost thaw and anchoring soil particles to prevent their removal by running water. Additionally, wildlife habitat interference and modification is reduced if a plant cover can be reestablished on surface disturbances. In areas where tourism and aesthetics are important, then plant-covered terrain is preferable to bare ground. There are many reasons why it is important to revegetate surface disturbances and this is not an exhaustive list. As a result, much effort is expended and expense incurred in attempts to reestablish plant cover on disturbances.

There is a debate about the best procedures in revegetation programs. Some maintain that native plant species are the best plants to use while others prefer agronomic taxa. Between these positions are those who want to mix the two and utilize qualities and properties of both groups of plants. Tests and trials continue and much remains to be studied and discussed. Many Vaartnou's paper provides a point of view in favour of the use of native species and he has included results of trials undertaken in the Canadian northwest. Lary Bliss and Nancy Grulke report on a number of studies carried out throughout the Queen Elizabeth Islands. Their paper provides insights into the results of experiments, some of which were initiated in the early 1970s. Both of these papers enhance our understanding of revegetation processes using native species.

The Potential of Native Populations of Grasses in Northern Revegetation

Manivalde Vaartnou
M. Vaartnou & Associates
11520 Kestrel Drive
Richmond, B.C.
V7E 4E2

Introduction

The purpose of this paper is to consider the recommendation of Thomas Berger (1977) for the use of native and naturalized grasses in northern revegetation. In 1977 a limited data base existed on the use of native cultivars in revegetation programs. Now, nine years later, this paper will address this deficiency and indicate the potential utility of some northern populations of grasses in northern revegetation work.

Reclamation refers to the return of the site to such a condition that it will be habitable to organisms originally present in approximately the same composition and density as before. It is acceptable if the site is made habitable to other organisms that closely approximate the original. Most environmentalists support this concept and recommend that, wherever possible, the original native species should be used in the reclamation process. This then would be *restoration*.

Native species are species indigenous to a given area. The term contrasts with "exotic species," which refers to species introduced directly or indirectly by man into any given area. A northern population, on the other hand, is a naturally occurring selection from within a species which may have a continental or circumpolar range. The population specific to a northern region such as Yukon Territory has evolved in response to its environment. Thus, when reference is made to "native species," most speakers or writers doubtless mean local populations of native species.

Native selections are usually the optimal for revegetation regardless of whether the objectives of revegetation pertain to aesthetic enhancement, improvement of wildlife habitat, agricultural development, or promotion of long-term ecosystem recovery. Native selections are assumed to be superior to exotic agronomic cultivars because

- they have adapted to the winter climate and will less often experience winterkill;

- they have adapted to the local photoperiod and thus have a greater reseeding potential;

- they will blend in easily with the landscape, thereby allowing a more harmonious end vista; and

- they often require less fertilization and allow for use of lower seeding rates.

Some grasses native to northern regions, such as polar grass — *Arctagrostis latifolia* (R.Br.) Griseb. — and bluejoint reedgrass — *Calamagrostis canadensis* (Michx.)

Beauv. — are known to have a slow early growth rate. These are poor candidates to provide initial erosion control, but are useful for long-term soil stabilization. Others, such as violet wheatgrass — *Agropyron violaceum* (Hornem) Lange — and fowl bluegrass — *Poa palustris* L. — which are among the primary invader species of disturbed areas, have a very rapid early growth rate and are extremely useful for initial erosion control. Both types of grasses are essential for successful revegetation while native legumes and other forbs also have a place in reclamation.

In order to illustrate the potential for native northern grasses in revegetation, results are presented from a revegetation trial site located in the Richardson Mountains of Yukon Territory and the seed production potential of northern populations of grasses is discussed.

Location and Methods

The location of the site is indicated in Figure 1. It is in the Richardson Mountains on the south side of the Dempster Highway where the highway intersects with the Yukon-Northwest Territories border. The aspect is level and the growth medium is the dark shale typical of borrow pits and overburden storage areas throughout the region. At the time of seeding vegetative cover was zero.

Twenty-one selections of grass seed were sown on 7 June 1979. Four were commonly used commercially available agronomic cultivars and the remainder were northern selections previously collected from northern British Columbia, northern Alberta, Yukon Territory or the Northwest Territories. Seed from the initial northern collections had previously been increased in northern Alberta.

The candidate grasses were hand-seeded to microsites in rows. At each microsite from five to ten seeds were placed within a 1 cm^2 area into a small depression and were then covered lightly with soil. Each entry was seeded to ten microsites per row. Row order was by random selection and spacing between rows and between microsites within each row was 1 m. All seeding was replicated three times. Previous soil sampling had indicated a deficiency of nitrogen and phosphorous, but sufficient potassium, for growth of grasses. Thus, the site was fertilized with 16-20-0 fertilizer at the rate of 80 kg/ha at the time of seeding. On 11 June 1981 an additional seven northern selections were seeded to the site. On this latter date no additional fertilizer applications were made.

The entries were evaluated for emergence in the year of seeding, subsequent survival, plant vigour as expressed through gross morphology and phenological development as expressed through production of seed. The emergence and survival evaluations consisted of an exact count of all possible microsites which had a live plant. The totals were then converted to percentages. Vigour ratings of 1 to 5 were subjectively assessed for each entry's development. The ratings were based upon the gross morphology and phenology of each entry as compared to its appearance in natural surroundings. Plant characters considered in this visual rating included leaf colour, leaf width and length, tillers, signs of disease, and rhizomatous development, if applicable. The numerical 1, 2, 3, and 4 values should generally be interpreted as corresponding to the words poor, fair, average, and strong, respectively. The 5 rating denotes successful phenological development of the entry. This was only used for entries which were rated at 4 for gross morphology and which also produced seed on a minimum of 50% of extant plants. The time frame available for development was considered in the early years of the program.

33

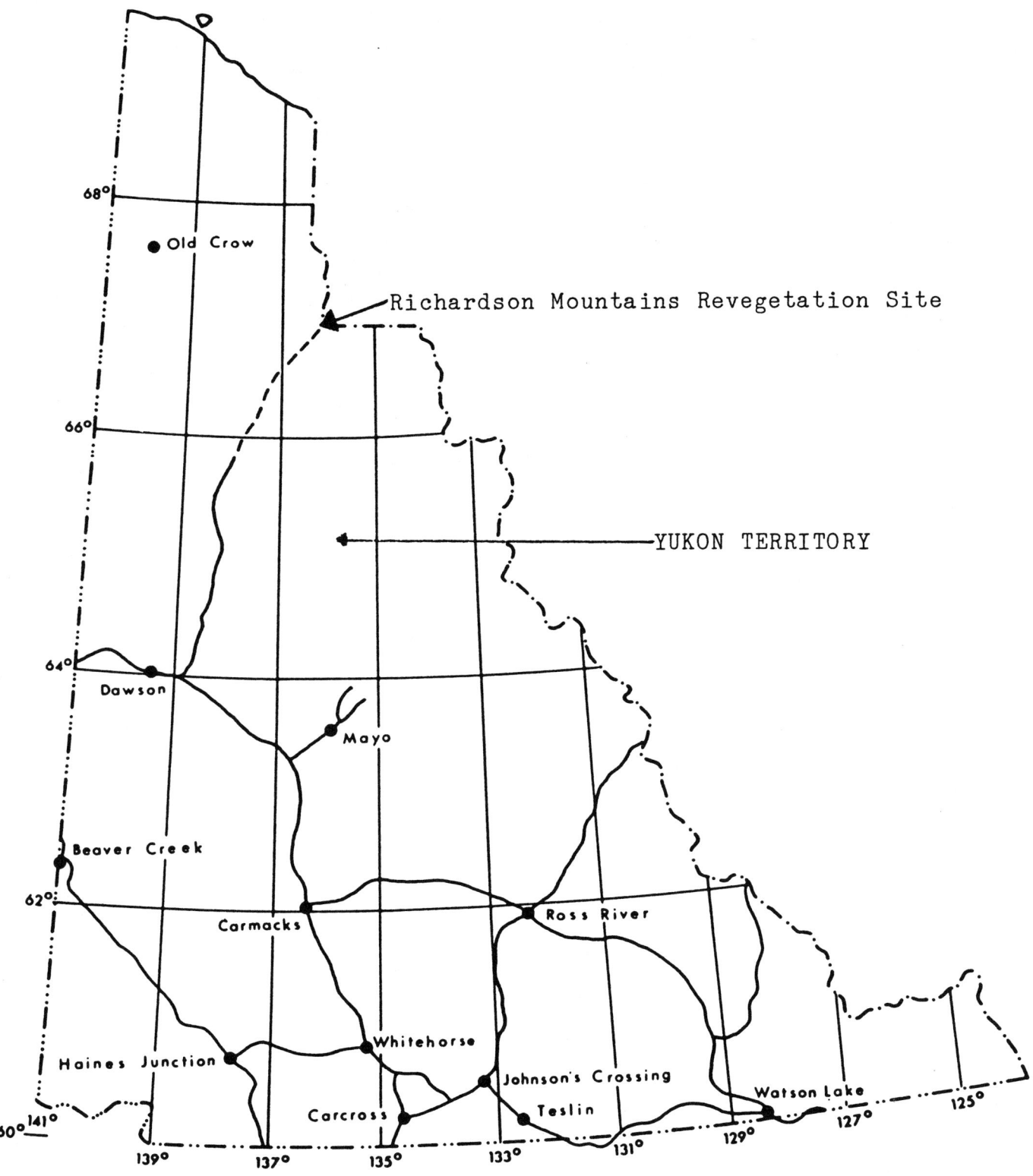

Figure 1. Location of the Richardson Mountains Revegetation Trial Site (adapted from Oswald and Senyk 1979).

By the time of the 1985 evaluation this was of minimal concern as even the most recently seeded entries had had five growing seasons to develop.

The seed production figures were based on an exact count of plants which produced seed. These totals were then converted to two percentage values. The first was the percentage of microsites containing plants which produced seed, in relation to all possible microsites, and the second was the percentage of existing live plants which produced seed.

Results

Emergence in the year of seeding (1979) was very successful as fourteen of seventeen northern selections and all four agronomic cultivars had plants in at least 93.3% of their microsites (Table 1). After one winter, the 1980 evaluation indicated that many entries were unsuited to this harsh environment as only 63.6% of the microsites contained live plants, in contrast to the original emergence of 93.0%. Plant mortality was near ubiquitous as the northern wheatgrasses (*Agropyron* spp.) decreased from 88.3 to 52.1%, the northern fescues (*Festuca* spp.) decreased from 100.0 to 86.7%, the northern bluegrasses (*Poa* spp.) decreased from 99.7 to 72.0% while the agronomic cultivars decreased from 99.2 to 47.5%. This downward trend continued in 1981 and by the time of the 1985 evaluation it was apparent that those selections which had high survival percentages

Table 1. Emergence and Survival of Grasses Sown at the Richardson Mountains Revegetation Trial Site on 7 June 1979

Species	Stock No.	Emergence (%) 1979	1980	Survival (%) 1981	1985
Agropyron cristatum	Fairway	96.7	10.0	0.0	0.0
Agropyron pauciflorum	159	30.0	16.3	10.0	10.0
Agropyron pauciflorum	Revenue	100.0	56.7	6.7	0.0
Agropyron riparium	209	100.0	86.7	76.7	20.0
Agropyron Smithii	9	100.0	86.7	36.7	0.0
Agropyron violaceum	103	100.0	66.7	53.3	33.3
Agrostis scabra	105	100.0	96.7	93.3	100.0
Alopecurus pratensis	5	100.0	70.0	60.0	30.0
Bromus inermis	Carlton	100.0	60.0	50.0	0.0
Bromus Pumpellianus	110	96.7	63.3	40.0	0.0
Deschampsia caespitosa	30	76.7	83.3	70.0	100.0
Festuca ovina	100	100.0	93.3	70.0	100.0
Festuca rubra	164	100.0	80.0	76.7	0.0
Festuca rubra	Boreal	100.0	63.3	70.0	0.0
Phleum pratense	17	93.3	16.7	10.0	0.0
Poa alpina	45	100.0	96.7	96.7	100.0
Poa compressa	177	96.7	23.3	20.0	0.0
Poa glauca	58	100.0	100.0	100.0	100.0
Poa palustris	74	96.7	56.7	40.0	0.0
Poa pratensis	178	100.0	83.3	70.0	53.3
Puccinellia Nuttalliana	195	66.7	26.7	36.7	0.0

and vigour ratings in 1981 continued to thrive while those which had experienced decreases in survival or vigour by 1981 had, in many cases, completely died out by 1985. The successful entries had increased through natural reseeding and, in most instances, were now approaching solids rows with little spacing between individual plants.

After seven growing seasons the most successful entries were tickle grass (*Agrostis scabra* Willd.), hairgrass (*Deschampsia caespitosa* (L.) Beauv.), sheep fescue (*Festuca ovina* L.), alpine bluegrass (*Poa alpina* L.) and glaucous bluegrass (*Poa glauca* M. Vahl.). These five entries were nearly solid rows in 1985. The other entries ranged from 53.3% survival to total failure. In all, eleven of the twenty-one selections seeded in 1979 had been eliminated from the site by 1985. These unsuccessful entries included seven northern selections and all four agronomic cultivars.

Over the length of the program plant vigour was usually highly correlated to emergence and survival percentages. In most cases, selections which had high survival percentages also received high vigour ratings. For instance, in 1979, eighteen of twenty-one entries received a vigour rating of 4 (Table 2). In 1980 the mean vigour of surviving plants decreased to 2.2 from 3.7 in 1979. However, the entries which still had high survival percentages were rated from 3 to 5 for vigour. In 1981, the mean vigour rating for surviving plants increased marginally as previously strong entries remained strong and weaker entries increased in

Table 2. Vigour of Grasses Sown at the Richardson Mountains Revegetation Trial Site on 7 June 1979

Species	Stock No.	Vigour (1-5)*			
		1979	1980	1981	1985
Agropyron cristatum	Fairway	4	1	—	—
Agropyron pauciflorum	159	1	1	1	1
Agropyron pauciflorum	Revenue	4	1	1	—
Agropyron riparium	209	4	2	3	1
Agropyron Smithii	9	2	1	3	—
Agropyron violaceum	103	3	1	1	1
Agrostis scabra	105	4	5	5	4
Alopecurus pratensis	5	4	2	3	2
Bromus inermis	Carlton	4	1	1	—
Bromus Pumpellianus	110	4	2	3	—
Deschampsia caespitosa	30	4	4	5	5
Festuca ovina	100	4	3	3	4
Festuca rubra	164	4	2	4	—
Phleum pratense	17	4	1	1	—
Poa compressa	177	4	2	1	—
Poa glauca	58	4	5	5	5
Poa palustris	74	4	1	1	—
Poa pratensis	178	4	3	3	2
Puccinellia Nuttalliana	195	4	1	3	—

* 1=poor 2=Fair 3=Average 4=Strong 5=Normal

vigour. However, this was somewhat deceptive, as the weaker entries suffered losses of their weakest plants, thereby increasing the mean vigour of survivors.

By 1985, the vigour ratings of the ten surviving entries had coalesced into two groups. Of the five entries with highest survival percentages, the hairgrass (*Deschampsia caespitosa*), alpine bluegrass (*Poa alpina*) and glaucous bluegrass (*Poa glauca*) were rated at the maximum of 5, while the tickle grass (*Agrostis scabra*) and sheep fescue (*Festuca ovine*) were rated at 4. No other entry seeded in 1979 was rated higher than 2 in 1985 (Table 2).

Production of seed, which is not expected for most species of grasses in the year of emergence, was limited in 1979, as only three selections of bluegrass (*Poa compressa* L., *P. glauca*, and *P. palustris*) and Nuttall's alkaligrass (*Puccinellia Nuttalliana* (Schult.) Hitchc.) produced seed on more than 40.0% of existing plants (Tables 3, 4). By 1980, seed production had increased as three entries — tickle grass (*Agrostis scabra*), alpine bluegrass (*Poa alpina*), and glaucous bluegrass (*Poa galuca*) — produced seed on more than 60.0% of plants while hairgrass (*Deschampsoa caespitosa*) produced seed on 48.0% of plants. Seed production continued to increase over time as the four entries with highest seed production percentages in 1981 all had seed on at least 86.2% of ex-

Table 3. Seed Production of Grasses Sown at the Richardson Mountains Revegetation Trial Site on 7 June 1979 in Relation to all Possible Microsites

Species	Stock No.	Seed Production (% of total microsites)			
		1979	1980	1981	1985
Agropyron cristatum	Fairway	0.0	0.0	0.0	0.0
Agropyron pauciflorum	159	0.0	3.3	0.0	0.0
Agropyron pauciflorum	Revenue	3.3	3.3	0.0	0.0
Agropyron riparium	209	0.0	0.0	10.0	0.0
Agropyron Smithii	9	0.0	0.0	0.0	0.0
Agropyron violaceum	103	0.0	0.0	0.0	10.0
Agrostis scabra	105	0.0	83.3	90.0	100.0
Alopecurus pratensis	5	3.3	6.7	20.0	10.0
Bromus inermis	Carlton	0.0	0.0	0.0	0.0
Bromus Pumpellianus	110	0.0	0.0	0.0	0.0
Deschampsia caespitosa	30	0.0	40.0	63.3	100.0
Festuca ovina	100	0.0	13.3	16.7	100.0
Festuca rubra	164	0.0	0.0	33.3	0.0
Festuca rubra	Boreal	0.0	0.0	13.3	0.0
Phleum pratense	17	10.0	0.0	0.0	0.0
Poa alpina	45	0.0	60.0	83.3	100.0
Poa compressa	177	43.3	6.7	0.0	0.0
Poa glauca	58	43.3	83.3	100.0	100.0
Poa palustris	74	46.7	3.3	0.0	0.0
Poa pratensis	178	0.0	16.7	13.3	0.0
Puccinellia Nuttalliana	195	30.0	3.3	16.7	0.0

Table 4. Seed Production of the Surviving Grass Plants Sown at the Richardson Mountains Revegetation Trial Site on 7 June 1979

Species	Stock No.	Seed Production (% of Extant Plants with Seed)			
		1979	1980	1981	1985
Agropyron cristatum	Fairway	0.0	0.0	—	—
Agropyron pauciflorum	159	0.0	20.0	0.0	0.0
Agropyron pauciflorum	Revenue	3.3	5.9	0.0	—
Agropyron riparium	209	0.0	0.0	13.0	0.0
Agropyron Smithii	9	0.0	0.0	0.0	—
Agropyron violaceum	103	0.0	0.0	0.0	33.3
Agrostis scabra	105	0.0	86.2	96.4	100.0
Alopecurus pratensis	5	3.3	9.5	33.3	33.3
Bromus inermis	Carlton	0.0	0.0	0.0	—
Bromus Pumpellianus	110	0.0	0.0	0.0	—
Deschampsia caespitosa	30	0.0	48.0	90.5	100.0
Festuca ovina	100	0.0	14.3	23.8	100.0
Festuca rubra	164	0.0	0.0	43.5	—
Festuca rubra	Boreal	0.0	0.0	19.0	—
Phleum pratense	17	10.7	0.0	0.0	—
Poa alpina	45	0.0	62.1	86.2	100.0
Poa compressa	177	44.8	28.6	0.0	—
Poa glauca	58	43.3	83.3	100.0	100.0
Poa palustris	74	48.3	5.9	0.0	—
Poa pratensis	178	0.0	24.0	19.0	0.0
Puccinellia Nuttalliana	195	45.0	12.5	45.5	—

isting plants. Also, by 1981, other selections such as meadow foxtail (*Alopecurus pratensis* L.), red fescue (*Festuca rubra* L.) and sheep fescue (*Festuca ovina*), had seed on at least 20.0% of their plants.

The 1985 evaluation indicated that seed production had continued to increase over the intervening years. Of the ten entries which had plants surviving from the 1979 seeding, the five with 100.00% survival in 1985 all had seed on each plant. These very successful entries were tickle grass (*Agrostis scabra*), hairgrass (*Descampsia caespitosa*), sheep fescue (*Festuca ovina*), alpine bluegrass (*Poa alpina*) and glaucous bluegrass (*Poa glauca*). The remaining entries from the 1979 seeding had seed production percentages ranging from 0.0 to 33.3%.

Of the seven selections seeded in 1981, the two with highest survival percentages in 1985 were polar grass (*Arctagrostis latifolia*) and a selection of glaucous bluegrass (*Poa glauca*). These selections, which originally had emergences of 56.7 and 63.3%, respectively, had increased to 100.0% by 1985 (Table 5). Plant vigour of the selections added in 1981 also correlated to survival as the polar grass and glaucous bluegrass discussed above were rated at 5 while the other surviving entries received vigour ratings ranging from 1 to 4 (Table 5). The successful additions — polar grass (*Arctagrostis latifolia*) and glaucous bluegrass (*Poa glauca*) — had seeds on each plant. Two entries with lower emergence percentages — bluejoint reedgrass (*Calamagrostis canadensis*) and arctic blue-

Table 5. Emergence, Survival, Vigour and Seed Production of the Grasses Sown at the Richardson Mountains Revegetation Trial Site on 11 June 1981

Species	Stock No.	Emergence (%) 1981	Survival (%) 1985	Vigour (1-5)*		Seed Production (%)			
						Total Microsites		Extant Plants	
				1981	1985	1981	1985	1981	1985
Agropyron pauciflorum	1980	90.0	43.3	1	1	0.0	10.0	0.0	23.1
Agropyron sp.	18	100.0	60.0	2	2	0.0	6.7	0.0	11.1
Arctagrostis latifolia	1980	56.7	100.0	1	5	0.0	100.0	0.0	100.0
Calamagrostis canadensis	1980	26.7	23.3	1	2	0.0	23.3	0.0	100.0
Festuca altaica	1980	56.7	0.0	1	—	0.0	0.0	0.0	—
Poa arctica	1980	20.0	50.0	1	4	0.0	50.0	0.0	100.0
Poa glauca	1980	63.3	100.0	1	5	0.0	100.0	0.0	100.0

*1=Poor   2=Fair   3=Average   4=Strong   5=Normal

grass (*Poa arctica*) — also produced seeds on each plant (Table 5).

Discussion

The data from the Richardson Mountains trial lead to three major conclusions regarding the long-term utility of the agronomic cultivars and northern selections seeded to the site. These are:

- The four agronomic cultivars included in this trial were not adapted to northern Yukon Territory.
- Most native selections were equally ill-adapted for survival in northern Yukon Territory.
- Five native northern selections — two bluegrasses (*Poa alpina* and *P. glauca*), hairgrass (*Deschampsia caespitsoa*), tickle grass (*Agrostis scabra*), and sheep fescue (*Festuca ovina*) — were adapted for long-term survival and can be considered suitable candidates for northern revegetation work.

In addition to the five very successful northern entries seeded in 1979, three of the selections added to the site in 1981 also had considerable promise and merit further study. These are polar grass (*Arctagrostis latifolia*), arctic bluegrass (*Poa arctica*) and the selection of glaucous bluegrass (*Poa glauca*) seeded in 1981.

The fact that the 1985 survival percentages of the very successful entries were, in most cases, higher than the original emergence percentages in 1979 is attributed to two separate factors. The minor part of these increases is a function of natural delayed germination but the major component is a result of the prolific seed production and emergence of progeny of the original plants. While the original planting allowed for 1 m gaps between plants to avoid confounding of results, the extremely high seed production rates of the successful entries have resulted in a near block plot effect on many sections of the site. Thus, the 1985 survival values are confounded but, from the practical aspect, the long-term durability of the successful entries has been even more dramatically emphasized.

Our data confirm the results of a shorter study done in the Macmillan Pass area of Yukon Territory. Brown (1985) used twelve northern selections, seven agronomic cultivars and one agronomic cultivar mixture as candidate grasses. In the years of seeding (1982), of the ten grasses which had the highest survival percentages, eight were agronomic cultivars. However, after four years (1985), of the ten grasses which now had the highest survival percentages, seven were northern selections.

Of some interest is the performance of the seven northern selections seeded at the Richardson Mountains trial site in 1981, two years after the site was fertilized. The data of Table 5 indicate initial emergence percentages ranging from 20.0 to 100.0%, with a collective mean emergence of 59.1% for these latter entries. This contrasts with the 91.6% collective emergence achieved by the northern selections seeded in 1979 (Table 1). Similarly, the northern selections seeded in 1979 had a collective first year vigour rating of 3.6, while the 1981 additions had a collective first year vigour rating of 1.1. Thus, while these figures are not directly comparable because different selections were seeded in the two years it does suggest that even native northern selections will benefit from the judicious application of appropriate fertilizer in the year of seeding. However, the 1985 data (Table 5) indicate that, of the six 1981 additions which had some surviving plants in 1985, three had achieved vigour of 4 or 5 by 1985. This suggests that these selections have adapted to both survival and growth in nutrient-limited conditions.

Two other considerations are of prime importance if native, northern selections are to be used in large-scale reclamation programs. These are the viability of seed and the economics of seed production. Since this study was undertaken only to assess survival, vigour, and amount of seed production in the natural environment, seed germination percentages could not be ascertained from these data. However, eight lots of seed of entries successful at various northern revegetation trials were sent to Canada Agriculture for germination testing. Of the eight, six can be deemed suitable for propagation as four were graded as eligible for Canada No. 1, one for Canada No. 2 and one for Canada No. 3. Consequently, while it is clear that each northern selection considered for seed increase should be tested for germination and purity of seed, it is equally clear that seed having high germination percentages can be obtained from northern stock material. Regardless of the germination percentages achieved in controlled test conditions plant emergence in the field will only be a small fraction of that achieved in the laboratory. Brown (1985), in his count study, found that only 19% of 37,200 seeds produced plants in the year of seeding. After four years, only 2.6% of the seeds had produced plants which still survived. From his data, one can infer that real-world environmental conditions will be the major factor which will determine the amount of seedling emergence, not a laboratory germination percentage difference of 1 to 5%. Also, when one considers that the cost of seed and fertilizer can be as small as 5% of the total northern reclamation package (D.M. Wishart pers. com.), it is easy to see that seeding rates could be increased without materially affecting the economics of northern reclamation.

The economics of northern seed production are also dependent upon choice of appropriate plants. Experiences with northern grass selections indicate that seed production difficulties and quanitity of seed produced vary tremendously. Problems associated with native seed use include poor seed production (Klebesadel 1974; Mitchell 1972), inappropriate growth form (Mitchell and McKendrick 1975) and seed harvesting difficulties (Klebesadel et al. 1962). Other commercial production problems may include seed shattering, low yields, hairs and awns, intermittent flowering, low fertility, and uneven flowering. Some of these problems can be overcome by genetic improvement or by developments in agricultural engineering technology (Walker *et al.* 1977). However, the most practical method is by appropriate selection, thereby minimizing the above difficulties by their avoidance.

The necessity for judicious choice of northern selections for economic seed production is illustrated by the wide ranges in kilograms of seed per hectare which are produced by three recently released cultivars selected from native Alaskan populations. Seed production of 300-1000 kg/ha is reported for Tundra bluegrass (*Poa glauca*) (Mitchell 1980a), 125-200 kg/ha for Alyeska polar grass (*Arctagrostis latifolia*) (Mitchell 1980b) and 20-35 kg/ha for Sourdough bluejoint reedgrass (*Calamagrostis canadensis*) (Mitchell 1980c).

Our own experience with northern seed production for the proposed Alaska Highway Natural Gas Pipeline reclamation program confirmed this situation. Many northern selections produce only minimal amounts of seed and should be avoided lest the cost of seed production drives the seed out of the commercial market. However, in northern Alberta, we achieved rates of production as high as 800 kg/ha with some wheatgrasses (*Agropyron* spp.) chosen for field-scale seed production. Also, bluegrass (*Poa* spp.) seed production rates ranging from 300-500 kg/ha were not at all uncommon. Thus, the seed could be produced in quantities which would allow for competitive marketing.

In conclusion, it is necessary to remember that the phrases "native species" and "northern ecotype" do not constitute an automatic panacea which guarantees revegetation success north of the 60th parallel. However, our experiences with northern selections of grasses indicate that viable seed can be obtained, economic quantities of seed production can be achieved and native northern selections will greatly outlive most agronomic cultivars. Therefore, while some agronomic cultivars developed from northern stock are likely to be useful in some revegetation situations north of the 60th parallel, the preferred strategy must be the inclusion of several native northern grasses in northern revegetation seed mixtures. Through the use of this strategy, the instant green-up followed by rapid vegetation disappearance, such as has occurred on the banks of the Alaska Highway in the proximity of Whitehorse, may be minimized in the future.

References

Berger, T.R. 1977. Northern Frontier Northern Homeland. The Report of the Mackenzie Valley Pipeline Inquiry. Volume Two. Terms and Conditions. — Ministry of Supply and Services. English Edition CP32-35/1977-2.

Brown, G. 1985. Results of Revegetation Experiments 1981-1985 Macmillan Pass, Yukon Territory. — Prepared for Northern Affairs Program, Yukon.

Klebesadel, L.J., Branton, C.I., and Koranda, J.J. 1962. Seed characteristics of bluejoint and techniques for threshing. — J. Range Man. 15: 227-229.

Klebesadel, L.J. 1974. Sweet holygrass, a potentially valuable ally. — Agroborealis 6: 17-20.

Mitchell, W.W. 1972. Adaptation of species and varieties of grasses for potential use in Alaska. — In: Proceedings of the Symposium on the Impact of Oil Resource Development on Northern Plant Communities. Occasional publication on northern life No. 1. Institute of Arctic Biology. University of Alaska, Fairbanks, pp. 2-6.

— 1980a. Registration of Tundra bluegrass. — Crop Sci. 20: 669.

— 1980b. Registration of Alyeska polargrass. — Crop Sci. 20: 671.

— 1980c. Registration of Sourdough bluejoint reedgrass. — Crop Sci. 20: 671.

Mitchell, W.W. and McKendrick, J.D. 1975. Tundra rehabilitation research: Prudhoe Bay and Palmer Research Centre, 1973-4 Summary Report to Alyeska Pipeline Service Co., ARCO, Canadian Arctic Gas Study Ltd., Exxon, Shell Oil and Union Oil. Institute of Agricultural Science, University of Alaska, Fairbanks.

Oswald, E.T. and Senyk, J.P. 1977. Ecoregions of Yukon Territory. Fisheries and Environment Canada. BC-X-64.

Walker, D., Sadasiviak, R.S., and Weijer, J. 1977. The utilization and genetic improvement of native Alberta grasses from the eastern slopes of the Rocky Mountains. Department of Genetics, University of Alberta, Edmonton.

Revegetation in the High Arctic: Its Role in Reclamation of Surface Disturbance

L.C. Bliss
Department of Botany
University of Washington
Seattle, WA 98195

N.E. Grulke
Systems Ecology Research Group
San Diego State University
San Diego, CA 92182

Introduction

The High Arctic of the Arctic Archipelago has a very different landscape, climate, and biology compared with the Low Arctic of the mainland. The landscape has low relief except for the eastern islands; there are few lakes except on Banks, Victoria, and Prince of Wales Islands; and in general habitats are less diverse than in the Low Arctic. The High Arctic (Bliss 1981) includes the Canadian Arctic Archipelago (except for southern Baffin Island), northeastern Keewatin District, and Northern Labrador. Large areas in the islands are essentially devoid of vegetation, consequently soil-frost action, and sheet and gully erosion are common features.

Mean annual temperature averages -11 to -15°C at most southern locations (Boothia and Melville Peninsulas) but only -16 to -19°C in the northern islands. Where anticyclonic activity is greatest, precipitation averages 80 to 140 mm but increases in the Eastern Arctic where cyclonic storms are more common (300 to 600 mm). Summer temperature ranges from 1 to 8°C for the daily July mean (Maxwell 1981). Accumulated degree days (>0°C) for the growing season range from 150 to 600 compared with 600 to 1200 in the Low Arctic. Depth of the active layer averages 20 to 80 cm which is greater than in the Low Arctic (20 to 60 cm) due to reduced plant cover, a minimal organic layer, and relatively well-drained soils.

Vascular plants provide only 5 to 25% cover in the polar semidesert landscapes. In the few areas of graminoid-moss tundra, vascular plants provide 30 to 50% cover (Muc 1977, Bliss and Svoboda 1984). This contrasts with vascular plant cover of 1 to 5% in the herb barrens of the polar deserts (Bliss et al. 1984). Cryptogams, especially crustose and foliose lichens provide 30 to 70% cover in many polar semidesert areas. These cryptogam surfaces play a central role in the establishment and maintenance of most vascular plant species (Sohlberg and Bliss 1984). Vascular plants tend to be smaller in size and have greatly reduced root systems compared with the same or closely related species in the Low Arctic (Bell and Bliss, 1978, Maesson et al. 1983, Grulke and Bliss 1985).

Vascular plants number about 350 species in the arctic islands with most islands having only 50 to 150 species. The graminoid, cushion, and rosette growth forms predominate with very few shrub species except in the southern islands. It is not surprising that the flora includes few if any species that can be called ruderal in this extreme environment. Human-disturbed tracts of land have only existed since the 1970s.

Observations and research on the growth forms and species that have some potential for reclamation in the Arctic Archipelago have extended from 1970 through 1986. Funding support has come from the Arctic Land Use Research Program (ALUR) Canadian Department of Indian and Northern Affairs, the National Research Council of Canada, the Boreal Institute for Northern Affairs, and the United States National Science Foundation grants (DEB 79-234260 and 81-07520). Logistical support was provided by the Polar Continental Shelf Project, Department of Energy and by Panarctic Oils Ltd. Field camp facilities were provided by Panarctic Oils, Ltd. and by the Arctic Institute of North America.

Landscape Observations of Plant Colonization

In the first survey (1970) of soil surface disturbance from petroleum exploration, *Papaver radicatum, Draba* spp., *Cerastium beeringianum, Oxyria digyna, Puccinellia angustata,* and *Saxifraga* spp. were reported as rapid invaders of denuded sites (Babb and Bliss 1974). However, it has been demonstrated from further observations on four of the Queen Elizabeth Islands that only graminoid species provided significant cover in many undisturbed plant communities. These included *Carex stans, C. membranaceae, Eriophorum triste* and *Dupontia fisheri* in moist to wet soils; *Luzula confusa, L. nivalis, Alopecurus alpinus, Phippsia algida,* and *Puccinellia vaginata* in upland, moist to dry soils. From this group of species *Carex stans* (Muc 1977), *Luzula confusa* (Addison and Bliss 1984), *Phippsia algida* and *Puccinellia vaginata* (Grulke 1983), and *Alopecurus alpinus* (Nosko 1984) were selected as representatives of their growth form for intensive study of their life history and ecophysiology.

Ecological studies, potential species

Carex stands Drej.

Carex stans or *C. aquatilis* Wahlenb. var. *stans* (Drej.) Boott (Porsild and Cody 1980) is a circumpolar species that dominates wetlands throughout the eastern High Arctic. While present in the northwestern islands, it is a minor component except on Melville Island. This rhizomatous species producers many tillers (1000 to 2000 shoots/m^2) and has the ability to slowly invade denuded soils. Vehicle tracks established in the mid-1960s took fifteen to twenty years to fill with *Carex stans* on the Truelove Lowland, Devon Island. A single pass of a tracked vehicle through a sedge-moss plant community on the North Sabine Bay Lowlands, Melville Island, in 1977 resulted in a 49% reduction in *Carex stans* over two years (Bliss 1979). Rhizomes had invaded the disturbed soils but were estimated to take ten to twenty years to provide a complete cover.

On average, 6 to 9% of the shoots flower per year but few if any of the seeds are viable. Seedlings of *Carex stans* are rare. Flowering intensity and shoot production were greater in warmer summers within plastic greenhouses where vapor pressure deficit was low and temperature averaged 7°C above ambient (Muc 1977). Individual shoots live five years and produce two to three leaves per year with the youngest leaves overwintering (Table 1). About 10% of the green tissue survives overwinter, which gives the plants an advantage for early growth as soon as snow melts. Shoot growth is completed in thirty to thirty-five days of the typical fifty to fifty-five-day growing season. Ratios of root:shoot biomass are 6:1 to 11:1 and for annual production 3:1 to 4:1. Thus roots live two to three times as long as shoots. These massive roots and rhizomes account for most of the accumulation of organic matter in these wetlands (Muc 1977).

Table 1. Comparison of characteristics of selected High Arctic graminoids.

Characteristic	Carex stans[1]	Luzula confusa[2]	Alopecurus alpinus[3]	Puccinellia vaginata[4]	Phippsia algida[5]
Growth form	upright	tufted	upright	prostrate	prostrate
Leaves produced annually	2-3	2	3-4	2	2
Leaves function (no. years)	1+	2	1+	1+	1+
Ramet age (years)	5-7	6-7	3-4	—	—
Genet age (years)	50-100+	90-130+	—	36+	24+
Root:shoot ratio (live)	6:1 to 11:1	0.3:1	1.3:1	2:1	0.8:1 to 2:1
Leaf water potential					
range MPa	-0.4 to -1.2	-0.2 to -0.7	-0.2 to -0.9	-1.2 to -2.6	-0.9 to -3.2
minimum MPa	-4.0	-1.2	-1.5	-2.9	-4.8

1. Addison 1977, Muc 1977. 2. Addison and Bliss, 1984. 3. Nosko 1984. 4. Grulke 1983. 5. Grulke 1983

While *Carex stans* is well adapted for growth in High Arctic wetlands, its ability to colonize denuded soils is limited due to both low sexual and vegetative reproduction and low tillering rates (Muc 1977). This species has low transpiration rates, yet leaf water potentials were below -1.0 MPa most of the summer and several times reached -3.0 to -4.0 MPa. These low potentials probably resulted from reduced cell membrane permeability in these cold soils (Levitt 1980).

Luzula confusa Lindebl.

Luzula confusa is a circumpolar, wide-ranging arctic-alpine species. It is one of the dominant species in the cryptogam-herb and moss-graminoid communities of the central and western polar semidesert landscapes (Bliss and Svoboda 1984). This species has a tufted growth form, with a large amount of standing dead material. Each tiller produces two leaves per year which function 2 years; thus there is a 50% yearly turnover of green tissue. On average a tiller lives 7 years and the average plant survives 90 to 130 years (Addison and Bliss 1984) (Table 1).

Flowering is common in this species yet no viable seed is produced (Bell and Bliss 1980). However, flowering stimulates rhizome branching. At these latitudes two consecutive warm and long summers may be required to produce abundant floral primordia and permit seed maturation (Addison and Bliss 1984). Roots and rhizomes account for only a small proportion of plant biomass (root:shoot ratio = 0.3:1).

Under field conditions of high soil water content, photosynthetic rates were sensitive to temperature increases above 3°C. Under laboratory conditions photosynthetic rates decreased with leaf water potentials below -0.7 MPa. However, field leaf water potential seldom dropped below -0.5 MPa in the cool, moist summer of 1974. This helps explain why this species is restricted to more mesic sites, especially where bryophytes are abundant (Addison and Bliss 1984, Bliss and Svoboda 1984).

Luzula confusa is well adapted to the High Arctic environment, but it has little ability to establish in denuded soils. Environmentally favorable periods for seedling establishment must occur occasionally, but this has not been observed from 1972 through 1984 in the northwestern islands.

Alopecurus alpinus J.E. Smith

Alopecurus alpinus is a circumpolar, wide-ranging arctic-alpine species that occurs in species-rich communities and in monoculture on soils derived from shale. This species has a very wide ecological amplitude, occurring from very wet to dry soils that include graminoid-moss meadows, cryptogam-herb, moss-graminoid, and graminoid barrens in polar semideserts and snowflush communities in polar barrens (Bliss and Svoboda 1984, Bliss *et al.* 1984). However, it does not occur in the upland dry soils of the polar barrens.

Alopecurus has an erect graminoid growth form, yet in dry soils plants become spreading and stunted. Shoots produce three to four leaves per summer and a typical shoot produces twelve leaves, indicating an average shoot age of four years (Nosko 1984) (Table 1). Only one leaf overwinters in this species (Bell and Bliss 1977). It is difficult to age these plants because of their rhizomatous growth.

In natural communities flowering intensity is low, 1 to 5%, but it increases significantly in surface-disturbed and nutrient-enriched sites (Table 2). Plant cover and flowering density differ greatly throughout the Arctic islands. Flowering from the preformed buds begins within one to two weeks of snowmelt and reaches a peak about four to five weeks after snowmelt (Nosko 1984). Although inflorescences and seeds are common in *Alopecurus*, no viable seeds or seedlings have been found (Bell and Bliss 1980, Nosko 1984). Seedling establishment is a rare event as it is with *Carex stans* and *Luzula confusa*.

Because of its rhizome development and rapid increase in growth with the addition of fertilizer (see below), this species has considerable potential for revegetation, provided that plants are in the immediate area or that plugs can be established.

Phippsia algida (Sol.) R. Br.

Phippsia algida is a widespread, circumpolar species, although it occurs as far south as the central Rocky Mountains near perennial snowbanks. It is a common species (2-6% cover) in mesic, polar, semidesert communities but also occurs in the drier graminoid barrens (Bliss and Svoboda 1984, Grulke 1983). Typical plant densities range from 150 to 400 plants/m^2 in the polar semidesert to over 1000 plants/m^2 in disturbed and nutrient-enriched sites (Grulke 1983).

Table 2. Plant cover and flowering density of *Alopecurus alpinus* on undisturbed and surface disturbed sites in the High Arctic.

Site	Degree days	Undisturbed Cover (%)	Undisturbed Flowering density (#/m^2)	Disturbed Cover (%)	Disturbed Flowering density (#/m^2)	Source
Rea Point, Melville Island	210	2.2	3.3	26.3	25.3	Bliss 1982
Truelove Lowland, Devon Island	331	12	7	—	144	Nosko 1984
Cape Abernethy, King Christian Island	160	2.3	1.8	—	—	Nosko 1984

Phippsia algida is a small, prostrate plant with ranked leaves 5-10 mm in length. It produces two leaves per culm per year, but with nitrogen fertilization leaf production often doubles (Grulke, unpublished data). Most of the leaves produced in a given year persist until midseason of the next year. Root to shoot ratio varies from 0.8:1 in seedlings to 2:1 in mature plants. In cold, dry years, root:shoot ratios are 1:1 in mature plants (Grulke 1983). Clumps of *Phippsia* may form in sites protected from the wind or in nitrogen-enriched sites due to seedling establishment near adult plants, although clumps never coalesce to form a vegetation mat, as does *Alopecurus alpinus*.

Plants live twenty-four years on average and begin to flower in about eighteen years. Approximately 25% of the population is mature enough to reproduce, but only one-third of this subpopulation produces inflorescences in any one year (Grulke and Bliss 1987). The number of plants induced to flower in a cold, dry year is one-half that in a warm, mesic year. The number of infloresences produced per plant (1.5) and the number of seeds produced per infloresence (22) is constant regardless of the environmental conditions during that year. Seed viability varies from 40% in a cold, dry year to 100% of seeds produced in a warm, mesic year (Grulke 1983). Flowering rates increase with phosphorus and iron fertilization and within small, plastic greenhouses. Seed germination in nutrient-enriched sites was twice that in control sites (Grulke, unpublished data).

This species naturally recolonizes disturbed sites. Although seedling recruitment from soil seed banks appears to be low, many viable seeds are produced most years. Seedling establishment of most species is restricted in bare soil by needle-ice formation and drying of surface soils. In *Phippsia*, the upper 1 cm of the root is corkscrew shaped, which hypothetically allows the roots to survive soil heave associated with freeze-thaw cycles without damage. All of these characteristics, in addition to same-year response to fertilization, indicate that this species is an excellent choice for revegetation of human-disturbed sites.

Puccinellia vaginata (Lge.) Fern. and Weath.

Puccinellia vaginata is restricted to Arctic North America, from the Mackenzie River Delta to Churchill, Manitoba, and north to eastern Greenland. It is a common species in graminoid barrens (1-3%) comprising 50% of the vascular plant cover (Grulke 1983). Typical density of *Puccinellia* is 7 plants/m^2 in these habitats.

Puccinellia vaginata is a medium-sized, prostrate plant with ranked leaves 10-20 mm in length. It produces two leaves per culm per year and most of the leaves produced in a given year persist until midseason of the next year. Root:shoot ratio is constant for all maturation stages and different environmental conditions (2:1). Plant size is greater in nutrient-enriched sites, but increased plant density is not apparent in these sites. Leaf production is not obviously enhanced by nutrient applications.

On average, plants live thirty-six years and begin flowering at twenty-seven years. Approximately 70% of the population is mature enough to reproduce and 60% of the reproductive-aged population produces inflorescences in a favorable year (Grulke and Bliss 1987). However, in a cold, dry year, only 1% of the reproductive-aged population is induced to flower. If induced to flower, the number of inflorescences produced per plant (1.5) and the number of seeds produced per inflorescence (18) is constant regardless of the environmental conditions during that year. Seed viability averages only 10%. Flowering rates and seed viability were higher in nutrient-enriched sites. Although this species is widespread in the drier graminoid barrens, it would not be effective in revegetation because of the low number of viable seeds produced (11 seeds/m^2), and because few seeds are produced in cold, short, or dry summers.

Experimental Plantings

King Christian and Devon Islands

In 1973 *Calamagrostis canadensis* and *Arctagrostis latifolia* seeds from north of Inuvik in the Mackenzie River Delta region (69°N) were planted in eight 1x2 m^2 plots at Cape Abernethy, King Christian Island, and Truelove Lowland, Devon Island (75°N). Seed of *Arctagrostis* from the Low Arctic was used because seed production is rare in the plants on the Truelove Lowland. *Calamagrostis* is not native to the Canadian Arctic Archipelago (Porsild and Cody 1980). These species were used to test the hypothesis that neither one would grow in the northwestern islands with their colder and shorter summers, but would survive in the eastern islands.

Seeds of each species were planted in separate plots at the rate of 20 kg/ha and fertilized at the rate of 300 kg N/ha and 1000 kg P/ha. At each site three plots of each species plus fertilizer and one control plot of each species minus fertilizer were established. The King Christian Island plots were sampled in 1975, 1977, and 1979. The Devon Island plots were sampled in 1974 and 1978. Laboratory tests of seed viability showed no germination in either species at 1° and 5°C. At 20°, germination in *A. latifolia* was 96% after six days and in *C. canadensis* it was 90% after eight days (Bell, unpublished data).

Arctagrostis and *Calamagrostis* survived for two or three years on King Christian Island, but by 1977 all plants had died. Overwinter mortality was high in these seedlings, probably the result of the lack of frost hardiness at freeze-up (Bell, unpublished data). In contrast, *Phippsia algida* seeded into these plots from adjacent areas and has increased its numbers over time, even after the removal of all plants in the 1977 sampling (Table 3). *Alopecurus alpinus* invaded vegetatively but did not reinvade following removal in 1977. Observations of these plots in 1985 showed an abundance of *Phippsia* but very few *Alopecurus* plants. These results demonstrate that the southern plantations of *Calamagrostis* and *Arctagrostis* are not adapted to survive on a typical island in the northwestern Queen Elizabeth Islands.

In contrast, these grasses grew better on the Truelove Lowland, Devon Island and they have remained viable through the 1986 growing season. *Arctagrostis* has maintained more plants and cover than *Calamagrostis* (Table 4). Fertilized plots developed a partial cover of bryophytes within the first two years and this probably reduced needle-ice formation. The unfertilized plots supported no seedlings of these species initially and they have remained devoid of vascular plants for thirteen years. Needle ice prevents the establishment of *Saxifraga oppositifolia*, *Juncus biglumis*, *Luzula confusa*, and *Oxyria digyna* but they have established from seed in the fertilized plots with their bryophyte cover (Figure 1).

Needle ice is a less important factor in coarse-textured soils on both Devon and King Christian Islands. However, in other studies on King Christian Island, it is apparent that needle ice is a key factor in seedling establishment and survival in bare soil, but much less so where lichens and bryophytes provide surface cover (Bell and Bliss 1980, Sohlberg and Bliss 1984, Grulke 1983).

Melville Island

In 1972 Panarctic Oils, Ltd. established revegetation plots at Rea Point, Melville Island to study the survival of seven agro-

Figure 1. Seeded plots of *Arctagrostis Latifolia* and *Calamagrostis canadensis* established in 1973, Truelove Lowland, Devon Island. Both species have maintained their populations in the fertilized plots but the unfertilized plots are still barren.

Table 3. Plant density (no./m^2) in seeded, fertilized plots and seeded, unfertilized plots, King Christian Island. Plots were seeded (20 kg/ha) and fertilized (300 kg N/ha, 1000 kg P/ha) in 1973 (means ± SD).

Species	1974 Fertilizer	No fertilizer	1975 Fertilizer	No fertilizer	1977* Fertilizer	No fertilizer	1979 Fertilizer	No fertilizer
Arctagrostis latifolia	0	0	37 ± 21	96 ± 68	0	0	0	0
Calamagrostis canadensis	0	0	9 ± 11	20 ± 8	0	0	0	0
Phippsia algida	0	0	14 ± 14	10 ± 8	10 ± 7	4 ± 1	169 ± 220	75 ± 77
Alopecurus alpinus	0	0	0	0	2 ± 1	1 ± 1	0	0

*All plants harvested.

Figure 2. *Alopecurus alpinus* as it occurs naturally on the sandy soils at Rea Pt., Melville Island (a) and its response to fertilizer after five years (b).

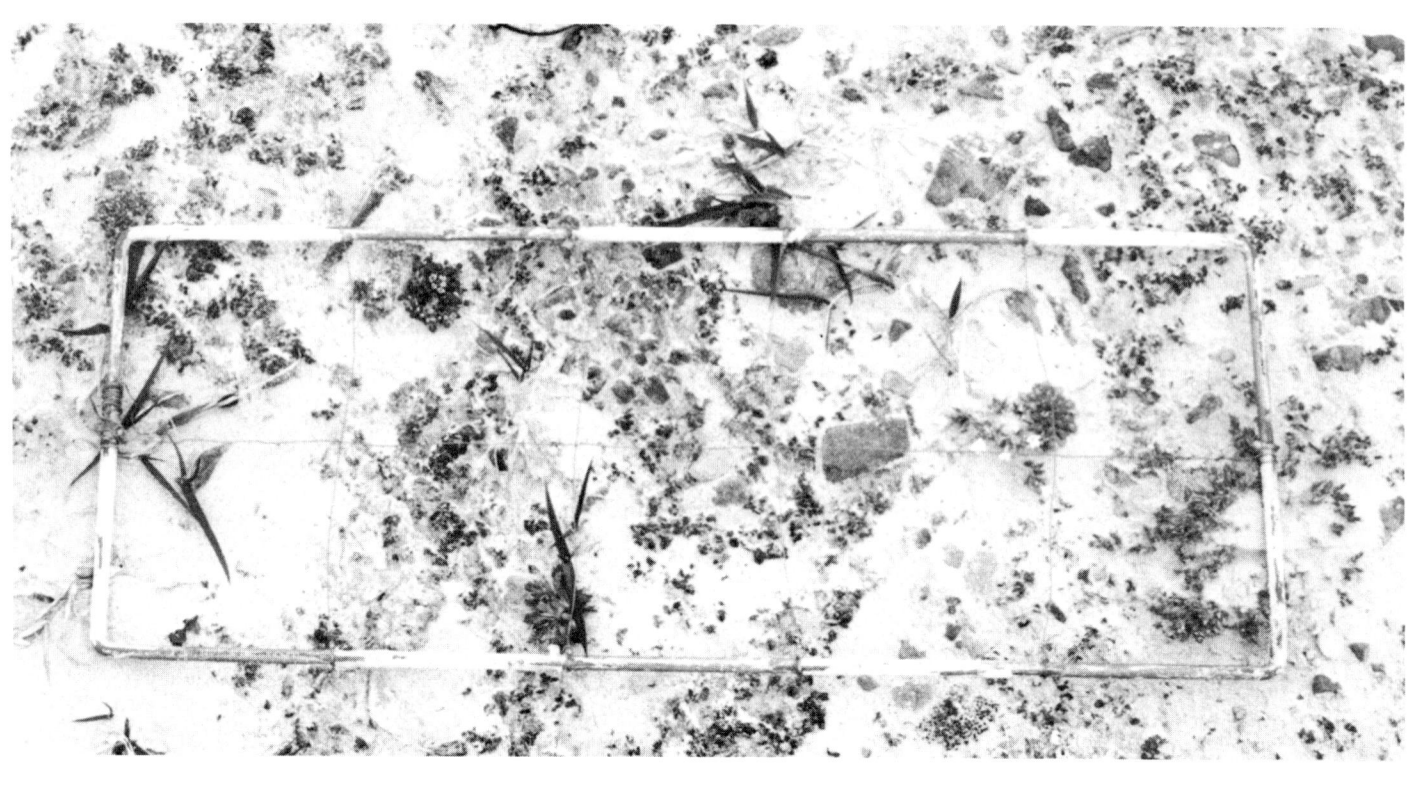

Table 4. Plant density (no./m²) in seeded, fertilized plots and seeded, unfertilized plots, Devon Island. Plots were seeded and fertilized (300 kg N/ha, 1000 kg P/ha) in 1973 (means ±SD).

Species	1974 Fertilized	1974 Unfertilized	1978 Fertilized	1978 Unfertilized
Arctagrostis latifolia	95 ± 36	31 ± 25	27 ± 11	0
Calamagrostis canadensis	11 ± 10	0	5 ± 6	0

nomic grasses in relation to fertilizer, seeding rate, seeding time (spring vs. fall), and seeding method (hand vs. hydroseeding). The sandy soils were tilled and the agronomic species planted (Griffing 1976, McGillivray 1976). Additional plots were established in 1974 and all were refertilized in 1976. These plots have been observed over the intervening years, and were sampled in July 1981. Sampling was confined to an adjacent undisturbed area (control) and to plots #20 and 42 of the original experiments (Figure 2a, b).

A few seeds of the agronomic species germinated in 1974 and 1975 but none survived one year. In contrast, the native grass *Alopecurus alpinus* survived the original tillage and established a sod within three years. These plots, now dominated by *Alopecurus*, have no lichens and a greatly reduced number of species. Plant cover and flower density of *Alopecurus* were significantly greater ($P < 0.01$) in the fertilized plots (Table 5). Living plant material, mostly net annual production of shoots as well as several years growth of roots, was significantly greater ($P < 0.01$) in the plots fertilized six and seven years earlier. Only *Alopecurus alpinus* was common to all plots and its production was again significantly greater in the former fertilized plots (Table 5), especially root and rhizome production.

These data demonstrate the tremendous response of *Alopecurus* to fertilizer and that the species can form a closed turf in less than five years. To our knowledge no other graminoid demonstrates this vegetative response to fertilizer in the High Arctic. Although the abundant flowering enhances tillering, it does not lead to seedling establishment. Over the ten years of observation, seedlings were never found nor were viable seeds produced.

Revegetation Program

From the above presentation it is evident that very few species have the potential for revegetation. *Phippsia algida* produces abundant seed, germination rates are high, and seedling establishment is successful, especially in bare soil with little competition from mosses, lichens, and vascular plants. However, this species grows slowly unless fertilized, and a closed turf does not develop.

Alopecurus alpinus flowers abundantly but produces no viable seed. However, rhizome development permits this species to spread and form a sod, provided fertilizer is added. Placement of sod plugs and fertilizer may be a feasible method of establishing a plant cover in some areas.

To test this approach, sod plugs of *Alopecurus* and *Phippsia* were established on a surface of drilling muds at the Hoodoo N-52 well site, Ellef Ringnes Island (78° 12'N, 99° 58'W) in 1985. The well was drilled the winter of 1981-82, cleaned up and studied in 1982. Although no plants had established on the drilling muds by 1985, seedlings of *Phippsia* were present on sandy soils (density 1.2 m²) near the well head and

Table 5. Plant cover, flower density, and plant growth on control (undisturbed) and tilled, fertilized plots, Rea Pt., Melville Island. Plots were fertilized in 1974, 1976, and sampled in 1981 (mean ± SE, n = 15 except plant growth n = 5).

Species	Plant Cover (%) Control	Fertilized Plots		Flower Density (no./m²) Control	Fertilized Plots		Plant Growth (g/m²) Control	Fertilized Plots	
	1	20	42	1	20	42	1	20	42
Alopecurus alpinus	2.2 ±0.6	26.3 ±6.8	25.3 ±6.5	3 ±6	218 ±118	116 ±38	11 ±5	447 ±100	353 ±176
Oxyria digyna	3.7 ±0.9	2.6 ±0.7	2.2 ±0.6	—	—	—	1.8 ±0.8	8.6 ±3.8	12.0 ±5.4
Total	13	5	8	—	—	—	—	—	—

within 50 m of undisturbed vegetation which included a few scattered plants of this species. The establishment of *Phippsia* at this site and its abundance on King Christian Island at the N-06 and D-18A well sites (Figure 3), demonstrate the ability of the species to invade and persist on disturbed surfaces. However, although large clumps of *Phippsia* may develop, they never form a vegetation mat even fifteen years after disturbance.

Based upon these studies and observations of numerous sites since 1971, the recommendation made in 1972 is still valid. The extremely slow growth rates of plants and the limited supply of seed indicates that reseeding scraped and rutted areas in the High Arctic does not seem feasible at this time (Bliss and Wein 1972). The application of fertilizer or spreading water from camp sumps will provide a higher nutrient level that will enhance the natural establishment of native species, especially *Phippsia*, on many upland sites.

Figure 3. Natural establishment of *Phippsia algida* at the D-18A well site on King Christian Island, twelve years after site cleanup.

References

Addison, P.A. 1977. Studies on evapotranspiration and energy budgets on Truelove Lowland.— In: Bliss, L.S. (ed.), Truelove Lowland, Devon Island, Canada: A High Arctic Ecosystem. Univ. Alberta Press, Edmonton, pp. 281-300.

Addison, P.A. and Bliss, L.C. 1984. Adaptations of *Luzula confusa* to the polar semi-desert environment.—Arctic 37: 121-132.

Babb, T.A. and Bliss, L.C. 1974. Effects of physical disturbance on arctic vegetation in the Queen Elizabeth Island.— J. Appl. Ecol. 11: 549-562.

Bell, K.L. and Bliss, L.C. 1977. Overwinter phenology of plants in a polar semi-desert. — Arctic 30: 118-121.

— 1978. Root growth in a polar semi-desert environment. — Can. J. Bot. 56: 2470-2490.

— 1980. Plant reproduction in a high arctic environment. — Arct. Alp. Res. 12: 1-10.

Bliss, L.C. 1979. Natural revegetation of summer and winter roads, airstrips, and well sites on Cameron and Melville Islands. — Panarctic Oils Ltd.

— 1981. North American and Scandinavian tundras and polar deserts. — In: Bliss, L.C., Heal, O.W., and Moore, J.J. (eds.),Tundra Ecosystems: A Comparative Analysis. Cambridge Univ. Press, New York, pp. 8-24.

— 1982. Revegetation in the High Arctic: Review of Grass Plots, Rea Pt., Melville Island. Report, Panarctic Oils Ltd. Calgary.

Bliss, L.C. and Svoboda, J. 1984. Plant communities and plant production in the western Queen Elizabeth Islands. — Holarctic Ecol. 7:325-344.

Bliss, L.C., Svoboda, J., and Bliss, D.I. 1984. Polar deserts, their plant cover and plant production in the Canadian High Arctic. — Holarctic Ecol. 7: 304-324.

Bliss, L.C. and Wein, R.W. 1972. Ecological problems associated with arctic oil and gas development. — In: Legett, R.F. and Macfarlane, I.C. (eds.), Proceedings, Canadian Northern Pipeline Research Conference. N.R.C.C. Technical Memorandum No, 104, Ottawa, pp. 62-77.

Griffing, T.C. 1975. An evaluation of Panarctic's experimental revegetation program in the arctic islands 1972-1975. — IEC, International Consultants Ltd., Vancouver.

Grulke, G.N. 1983. Comparative morphology, ecophysiology and life history characteristics of two high arctic grasses, N.W.T. —Ph.D. thesis, University of Washington, Seattle.

Grulke, N.E. and Bliss, L.C. 1985. Growth forms, carbon allocation, and reproductive patterns of high arctic saxifrages. — Arct. Alp. Res 17: 241-250.

— 1987. Comparative life history characteristics of two high arctic grasses. — Ecology (submitted).

Levitt, J. 1980. Responses of Plants to Environmental Stresses. Academic Press, London.

Maessen, O., Freedman, B., Nams, M.L.N., and Svoboda, J. 1983. Resource allocation in high-arctic vascular plants of differing growth form. — Can. J. Bot. 61: 1680-1691.

Maxwell, J.B. 1981. Climatic regions of the arctic islands. — Arctic 34: 225-240.

McGillivray, A.H. 1976. Panarctic Oils Ltd.

revegetation project five year progress report. Haddock's Ltd., Calgary.

Muc, M. 1977. Ecology and primary production of the Truelove Lowland sedge-moss meadow communities. — In: Bliss, L.C. (ed.), Truelove Lowland, Devon Island, Canada: A High Arctic Ecosystem. Univ. Alberta Press, Edmonton, pp. 157-184.

Nosko, P. 1984. A comparative study of plant adaptations of *Alopecurus alpinus* in the Canadian High Arctic. Ph.D. thesis, Univ. Alberta, Edmonton.

Porsild, A.E. and Cody, W.J. 1980. Vascular Plants of Continental Northwest Territories, Canada. — Canadian National Museum of Natural Science, Ottawa.

Sohlberg, E. and Bliss, L.C. 1984. Microscale pattern of vascular plant distribution in two high arctic communities. — Can. J. Bot. 62: 2033-2042.

Ecosystematic Research
Peter Kershaw
Editor

Previous papers in this volume show that theory and practice have progressed from *ad hoc* decisions made in an office, rather than on-site, about the most appropriate mitigation or reclamation practices. However, it can be argued that there is a need for further work and, in particular, the integration of results within a broader ecosystem perspective. Seldom has it been possible, given time and fiscal constraints, to test specific mitigation and reclamation practices which consider better engineering or surface treatments not in isolation, but as part of an integrated, multidimensional and functioning system. Furthermore, it is seldom possible to evaluate performance of these practices over time frames greater than a few years. Monitoring and logistic costs attendant to such efforts make it difficult to justify this type of a project to industry or government bodies. With the exception of studies conducted within the IBP few research efforts have been able to place disturbance studies within the broader context of the ecosystems affected. This is not to say that broad-based ecological studies have failed to consider other environmental aspects that contribute to the total picture on mitigation and reclamation. However, studies specifically designed to quantify ecosystem components and consider the implications of specific reclamation practices within this context have not been attempted. The following paper is an outline of such a study. It is being conducted within a subarctic, upland *Picea mariana* forest but the approach should be transferrable to any ecosystem.

The Use of Controlled Surface Disturbances in the Testing of Reclamation Treatments in the Subarctic

G.P. Kershaw
Department of Geography
University of Alberta
Edmonton, AB T6G 2H4

Introduction

North America contains a broad belt of open-canopied spruce forests poleward of the boreal forest or taiga, extending from the northwest in Alaska to central Labrador. In Eurasia a similar zone is found from the coast of Norway east through the Soviet Union to the Sea of Okhotsk and the Bering Sea (Tuhkanen 1984). This subarctic zone includes the tundra forest and the open woodland forest types of Rowe (Payette 1983). Variations in definitions of northern forest types make it difficult to provide an estimate of the areal extent of the subarctic region. However, taiga systems have been estimated to cover 10.09 million km² on a circumpolar basis, of which the northern portion (presumably the Subarctic) is 3.34 million km² (Bliss et al. 1981). In Canada 760,000 km² has been classified as Subarctic (Zoltai et al. 1987), or approximately 7.6 percent of the country (larger than the province of Alberta). With current and future developments in this spatially significant area of the northern hemisphere it is expected that the number of man-induced disturbances will increase.

At present a number of major development projects have occurred within the subarctic zone in North America — TAPS, the Norman Wells project, the Dempster Highway, and numerous smaller developments. Furthermore, there are a number of major and minor proposed developments that will also affect the Subarctic — the Mackenzie Highway, the Arctic Pilot Project, Phase two of the James Bay Hydro-electric Project, Polar Gas (a pipeline through the Mackenzie Valley), and others.

In Canada's North the Territorial Land Use Regulations stipulate that the permittee "shall ... restore the permit area as nearly as possible to the same condition as it was prior to the commencement of the land use operation" (Territorial Lands Act, SOR.77-210 4 March 1977). This requirement is difficult to implement in the Subarctic, where variability in short- or long-term time frames affect reclamation processes. Furthermore, few studies have been conducted to develop predictions as to the efficacy of certain reclamation practices (Younkin and Martens 1985). Often, it is by practice rather than pretesting that programs are developed and effected (Wishart, this volume). We have not had the luxury of *time* when attempting to provide answers to specific reclamation questions. *Time* to investigate the topic and then provide reliable answers before development proceeds. *Time* to observe test sites over the long term in order to accurately characterize ecosystem responses to man-induced disturbances within the biologically extreme and variable conditions that prevail in the North.

Subarctic ecosystems exist in environments

of extremes where *average* as a concept may often be inappropriate from abiotic and biotic perspectives. Perhaps the most important environmental control is exerted by climate. Within year and between year climatic variability in the Subarctic is probably greater than most other biomes in Canada. For example, variations in such a key environmental factor as air temperature have the greatest magnitude of any region in Canada (Statistics Canada 1986). Furthermore, annual deviations from the climatic normals have been among the greatest in Arctic Canada (Higuchi et al. 1986, Higuchi and Etkin 1987). This condition has significant implications for such critical environmental components as soil nutrient availability and active layer depth, growing season length, growing-degree-day totals, or snow depth and duration. This high degree of variability in climate, geomorphic processes, biological function, etc. makes it extremely difficult to make reliable long-term predictions (more than five years) of biological or physical environmental components. This is especially true when predictions are only based upon short-term studies.

In circumpolar northern lands, international co-operation on ecosystematic studies — through the International Biological Program (IBP) and Man and the Biosphere (MAB) initiatives — has been very productive in providing answers to questions on system function (Worthington 1975). However, in the area of ecosystem response to man-induced disturbances there has, at best, been only *ad hoc* collaboration and little commonality of approach.

Recently the International Geosphere-Biosphere Program (IGP) has been created by the International Council of Scientific Unions (ICSU). Its objective is "to describe and understand the interactive physical, chemical and biological processes that regulate the total earth system, the unique environment it provides for life, the changes that are occurring in that system, and the manner by which these changes are influenced by human actions" (Malone 1986). In agreement with this thrust is the Royal Society of Canada with its assertion that "the time has come to attack the major problem of changes in the geosphere and biosphere and their relationships, particularly as they involve human activities that can be modified to preserve and improve human welfare" (Royal Society of Canada 1985). Unlike IBP and MAB, this new international research program is directly concerned with human impacts on the global systems and mitigation measures. In Canada a number of topics have been identified. Changes in the boreal forest, permafrost dynamics, the 500-year record of change, soil degradation, erosion, biomass, hydrogeology, and the albedo of Canada are five of the eleven topics deemed appropriate for detailed consideration.

It is within this context that the Studies of the Environmental Effects of Disturbances in the Subarctic (SEEDS) project was developed 10 km north of Fort Norman, N.W.T. (Figure 1). It has the following objectives:

- to conduct a detailed monitoring of experimental rehabilitation treatments designed to manipulate disturbances initiated within a Subarctic *Picea mariana* ecosystem that were themselves a simulation of a transport corridor,
- to test, under controlled field conditions, several rehabilitation/revegetation treatments to determine their relative merits, and
- to develop a preliminary computer model that simulates the site conditions (Subarctic Environmental Response Model — SERM) and test it with the data collected over the duration of the project in order to improve on its predictive capabilities.

A number of key ecosystem components have been identified and studies designed to

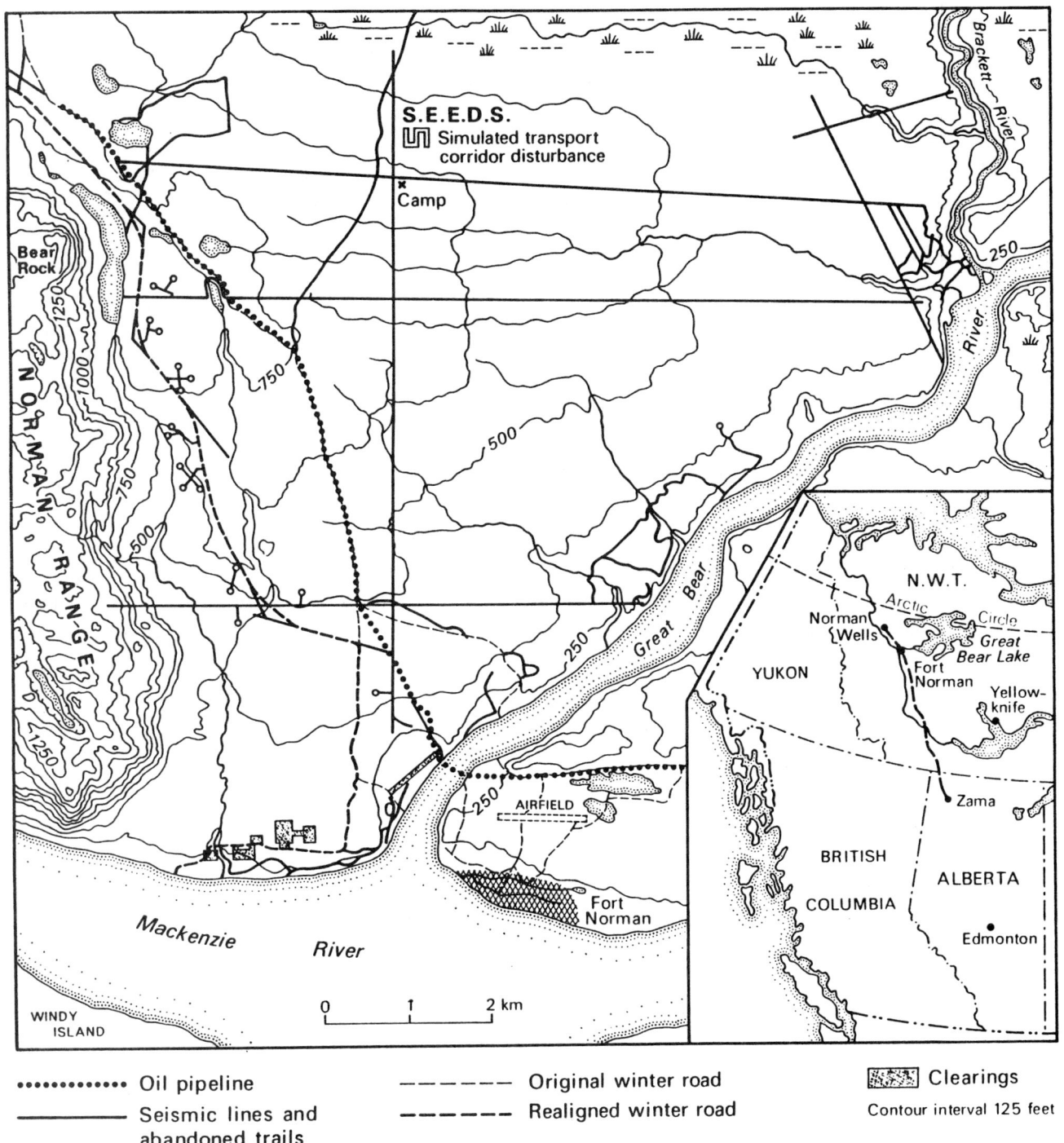

Figure 1 Location of the simulated transport corridor disturbance installed for the project – Studies of the Environmental Effects of Disturbances in the Subarctic (SEEDS), Fort Norman, N.W.T.

provide an integrated approach to the establishment of baseline data and responses to the controlled environmental-manipulation experiments have been initiated (Figure 2).

Project Overview

In order to assess the performance and success of test reclamation treatments it is desirable to have predisturbance data for many of the ecosystem variables. Additionally, in order to separate the ecosystem responses to the disturbances from the normal, natural changes inherent in the ecosystem it is essential to maintain a measurement program in the undisturbed adjacent areas (control) as well. In the case of the study at Fort Norman it was not possible to conduct a monitoring program within the entire area to be affected and so disturbances were initiated in certain areas while other sites were monitored for a year and then subjected to a series of disturbances. Furthermore, the scope and extent of reclamation treatments continues to expand with the addition of new test treatments in each of the operating years since 1985. However, control stands are maintained and left in an undisturbed state.

Reclamation Experiments

At the Fort Norman site a 25-m-wide simulated transport corridor was cleared over a two-year period (Figure 3). The north-south oriented corridors were identified as Right-Of-Way (ROW) 1, 2, and 3 from west to east. In ROW 1 a set of four trench segments was installed in 1985 with approximately 10 m of untrenched buffer between each (Figure 3). In ROW 3 a complete trench was created in 1985, including the east-west link. ROW 2 was installed in 1986 with its trench. Within the trench, alternating 50 m stretches were backfilled with slash prior to the soil backfilling operation (Figures 2 and 3). The slash was intended to provide a biodegradable insulative material. Trenching was done to the base of the active layer — an average of 50.7 cm (n=129, SD=14.6) — and the organic surface layer, including the moss and lichen layer, was mixed back with the mineral soil. This corridor was constructed to simulate those created by seismic, winter road, electrical transmission line, and pipeline rights-of-way. In addition, the trench was a simulation of a buried pipeline or any surface disturbance resulting in partial removal of surface organics or excavation and backfilling.

A number of test reclamation treatments were installed along the trench. They included combinations of seeding with agronomic or native species mixes or shrub whip plantings, and insulated or uninsulated backfill (Figure 4).

Abiotic Ecosystem Components

Microclimate

The initial microclimate network employing automated dataloggers was established in 1985 with the installation of one multi-sensored station. A network with control, cleared right-of-way, and trenched sites has subsequently been established with soil/permafrost and atmosphere parameters being measured (Figure 3). The number of sensors varies seasonally from a high of 170 (including a number of types — Figure 2) in the thaw season to 127 in winter.

Soil/permafrost temperature data are taken to a maximum of 225 cm with as many as six sensors in one location, positioned at various depths. There are a number of sensor depths common to many of the sites (2.5, 10, 50 and 200 cm below the moss surface). Additionally, surface temperatures, +10, +50, and +150 cm air temperatures are taken at five locations on the study site. Relative humidity at +150 cm; wind speed at +150 and +300 cm; net radiation (summer) at +150 and +400 cm and evapotranspiration (summer)

Figure 2. Ecosystem Components

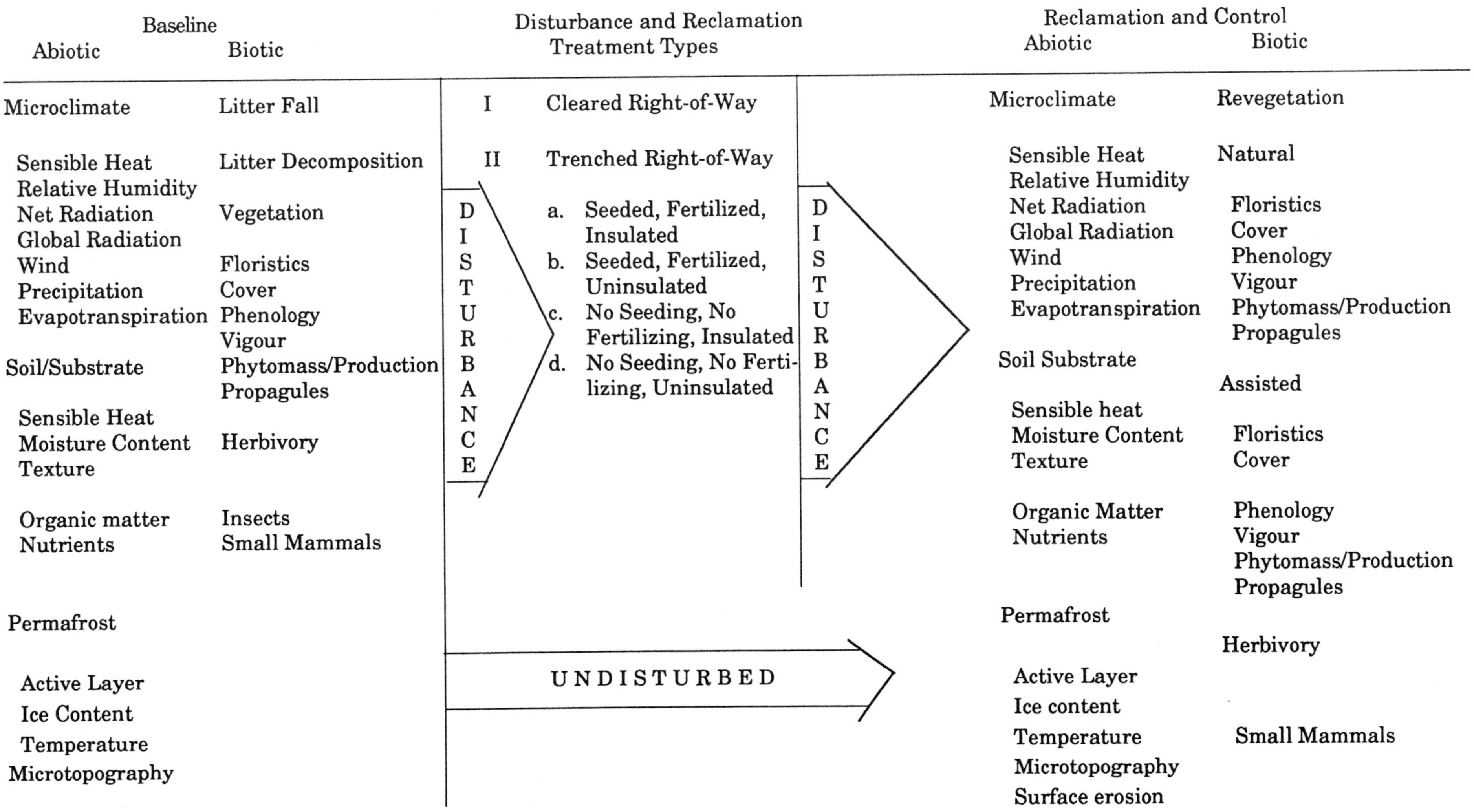

	Baseline		Disturbance and Reclamation Treatment Types	Reclamation and Control	
Abiotic	Biotic			Abiotic	Biotic
Microclimate	Litter Fall Litter Decomposition	I	Cleared Right-of-Way	Microclimate	Revegetation
Sensible Heat Relative Humidity Net Radiation Global Radiation Wind Precipitation Evapotranspiration	Vegetation Floristics Cover Phenology Vigour Phytomass/Production Propagules	II	Trenched Right-of-Way a. Seeded, Fertilized, Insulated b. Seeded, Fertilized, Uninsulated c. No Seeding, No Fertilizing, Insulated d. No Seeding, No Fertilizing, Uninsulated	Sensible Heat Relative Humidity Net Radiation Global Radiation Wind Precipitation Evapotranspiration	Natural Floristics Cover Phenology Vigour Phytomass/Production Propagules
Soil/Substrate	Herbivory		DISTURBANCE	Soil Substrate	Assisted
Sensible Heat Moisture Content Texture				Sensible heat Moisture Content Texture	Floristics Cover Phenology Vigour Phytomass/Production Propagules
Organic matter Nutrients	Insects Small Mammals		UNDISTURBED	Organic Matter Nutrients	Herbivory
Permafrost				Permafrost	Small Mammals
Active Layer Ice Content Temperature Microtopography				Active Layer Ice content Temperature Microtopography Surface erosion	

Figure 3. Abiotic ecosystem components and sampling locations; SEEDS, Fort Norman, NWT — 1985 and 1986

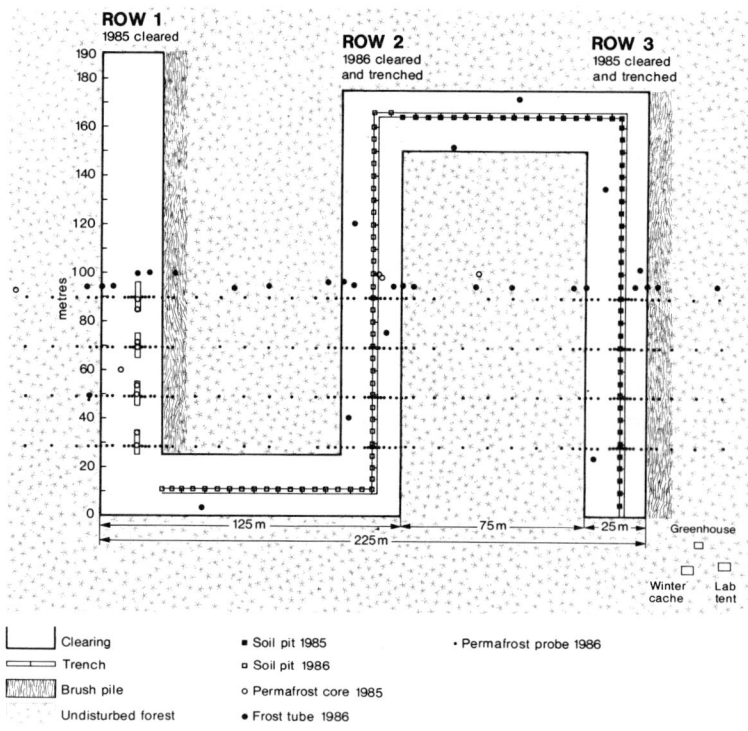

Figure 4. Biotic ecosystem components and sampling locations; SEEDS, Fort Norman, NWT — 1985 and 1986

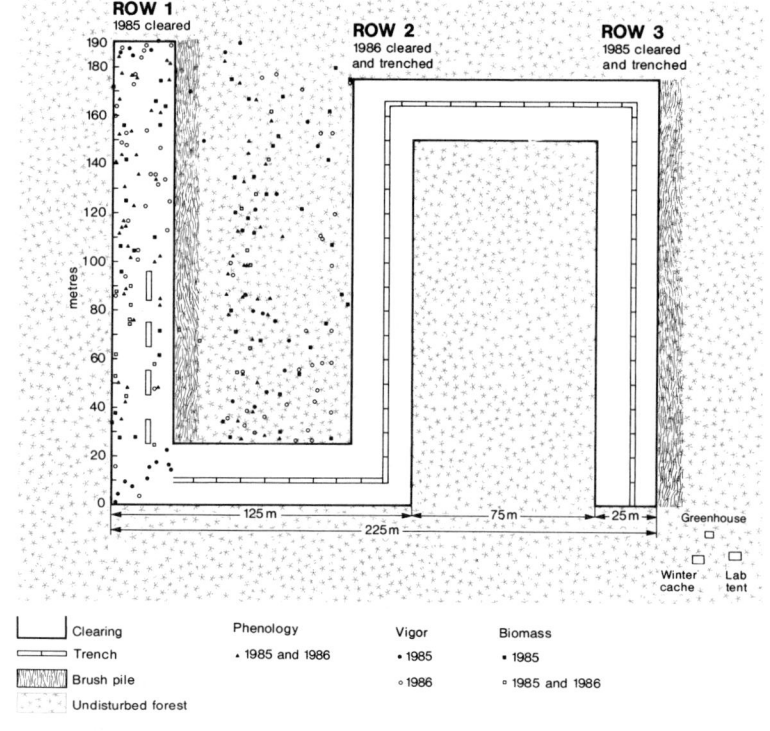

are all monitored at multiple locations on the site (Figure 2).

Soil/Substrate

Predisturbance soil sampling was conducted at 5 m intervals along the simulated pipeline trench, at the time of right-of-way clearing and trenching (Figure 3). These studies were initiated in 1985 with the collection of over 380 samples from 49 soil pits (Kershaw and Evans 1986). A further 310 samples were collected in 1986 from 51 soil pits (Evans *et al.* 1988). Data to be generated will include moisture content, organic matter content, texture, and macro- and micronutrient content. A set of soil temperature sensors were installed in 1985 and data on temperature profiles are currently being collected on a twelve-month basis by automated dataloggers.

Permafrost

Permafrost studies rely on four sources of data. First, there are 104 cores which extend 1.3 to 2.25 m below the moss surface. These core holes were sunk to install thermocouple strings, frost tubes, and pin frame anchor posts. This network (Figure 3) will be extended to provide a more comprehensive system for future studies. Material that was removed during this network installation will be analyzed in the lab, permitting quantification of ice contents and texture, with a resolution of from 5 to 15 cm depths (the coring depths between sample removal) (Evans *et al.* 1988, Kershaw and Evans 1986).

The second source of permafrost data is derived from information compiled by the microclimate instruments and associated thermocouple network (Table 3). Data are collected continuously from disturbed, reclamation test treatments, and undisturbed sites from depths down to 2 m below the permafrost table.

The third source of information on the permafrost is the frost tubes that have been inserted in certain core holes (Figure 2). Approximately thirty will eventually be installed and monitored, providing data on the rate and thickness of thaw and freezeback in the active layer.

The fourth source of permafrost information is available from the weekly thaw-depth surveys conducted along transects that extend through the cleared rights-of-way from undisturbed forest (Figure 3).

Topography

In order to quantify the subsidence occurring as a result of thermokarst processes, a number of topographic surveys have been initiated. The intermediate-scale surface topography was surveyed throughout the cleared rights-of-way and topographic maps will be produced to spatially define the situation at the time of tree clearing. This mesoscale of mapping will serve to identify the drainage systems that operate during spring runoff and to define the scale of subsidence associated with the simulated pipeline trench.

With microscale relief changes occurring within the trench, the resolution provided by the mesoscale mapping is inadequate. Consequently a number of pin-frame measurement sites will be installed to quantify these topographic changes. There will be installations on the trench; in the cleared rights-of-way and in the undisturbed forest. Using this method it will be possible to measure the extent and magnitude of heave, subsidence, and rill erosion. Over the next few years any surface elevation changes resulting from either thermal erosion or thermal subsidence will be quantified at these locations. It is anticipated that a resolution of 0.5cm will be possible for the microscale measurements.

Hydrology

Surface runoff is restricted to the snowmelt period and rarely during torrential rainfall events. A system of gauges will be installed and monitoring of discharges will be accomplished during the thaw season (Steer and Woo 1983).

Subsurface flow within the active layer will be monitored during the thaw season. This study will provide data necessary for an evaluation of the extent of soil heat flux associated with soil water. Runoff which occurs mainly in the snowmelt period can contribute to the erosion of soil along the trench.

Biotic Ecosystem Components

Litter fall and decomposition

Within the undisturbed area and ROW 3, sixty 50x25 cm litter traps were installed flush with the moss surface. The trays were lined with a fine-meshed nylon filter to prevent organic matter from being washed through the drain holes in the tray bottom. These trays were emptied on a monthly basis during the snow-free period and the total amount of collected material was weighed and separated into components based upon species.

Nylon packets containing litter were deployed within the undisturbed and disturbed areas to determine the rate of decomposition for eleven different types of litter. Included were leaf material only from the dominant evergreen shrub — *Ledum groenlandicum*; three deciduous shrubs — *Arctostaphylos rubra*, *Salix arbusculoides* and *S. myrtillifolia*; an evergreen tree — *Picea mariana*; a grass — *Arctagrostis latifolia*; and a sedge *Carex membranacea*.

Vegetation and Floristics

A number of studies on the site's vegetation were initiated. Quadrats 1x1m in size were randomly selected within the control and ROW 1 area. These sites were roughly located and permanently marked for resurvey. Approximately half of the sixty quadrats in ROW 1 and in the control area were phenology plots and the other half were biomass plots harvested in 1985 and 1986 (Figure 4). Complete floristic and species' cover were assessed within each of these plots in 1985.

The 125 quadrats randomly selected for study on the site were inventoried for species composition. A voucher collection of all species of vascular and nonvascular plants was made and a flora was compiled for each quadrat, i.e., all the phenology, vigour, and biomass plots (Figure 4).

Plant Cover, Phenology and Vigour

The cover of each taxon was visually estimated within the same 125 1x1m quadrats in which floristics were described. All structural levels were quantitatively assessed.

Within thirty-eight ROW 1 and thirty control quadrats plant phenotypic states were monitored throughout the growing season (Figure 4). In 1985 fourteen surveys were completed and in 1986 twenty surveys were done. Vegetative and reproductive phenophases were categorized (Ross and LaRoi 1984) and the first occurrence of each was noted for every vascular plant species in each quadrat. The quadrats are permanent and therefore the year-to-year variations in phenotypic states can be compared for each species on a quadrat basis as well as a study basis.

Two evergreen shrubs — *Ledum groenlandicum* and *Empetrum nigrum*; four deciduous shrubs — *Salix myrtillifolia*, *S. arbusculoides*, *Arctostaphylos rubra* and *Vaccinium uliginosum*; two sedges — *Carex membranacea* and *C. vaginata*; and a grass — *Arctagrostis latifolia* were selected for analysis of vigour. These eleven species form the dominant understory cover and biomass of the vascular plants. Sampling points were randomly selected and the above ground portion of the closest individual to each point was removed. Each specimen was divided into component parts and these were measured to determine dimensions (e.g. leaf width, length of underground shoots, etc.). Data are collected at the end of each growing season.

Shrub Whip Plantings

In the spring of 1986 hardwood cuttings were collected from shrubs that were successful colonizers on local seismic lines: *Salix arbusculoides, S. glauca, Vaccinium uliginosum, Betula glandulosa, Alnus crispa* and *Potentilla fruticosa*. For each species, two populations were sampled; plants located on a fourteen-year-old seismic line (disturbed site), and plants located in a *Picea mariana* stand that has remained unburned for 200 years (the control). The two population samples will be monitored for ecotypic responses. All cuttings were trimmed to whips of 10 cm in length and 150 whips/ species were planted at 25 cm intervals in otherwise untreated sections of the trench. One half of all the whips came from the seismic population and one half from the control population.

Whip cover and biomass were also assessed at the end of the growing season for a subsample (n=40) of each species. These data will serve as a temporal benchmark for comparison with collections at the end of subsequent growing seasons. In the spring of each year, percent overwinter survival will be assessed for all species.

Resprouting

Prior to clear-cutting ROW 2, 150 *Salix arbusculoides* were randomly selected and cut to ground level. This was done so that resprouting from the root mass could be quantified to determine what controls the nature and amount of resprouting following a clearing operation. *S. arbusculoides* was the dominant erect shrub in the stand and the only species that was present in large enough numbers to be sampled in this fashion. All above ground portions were separated into the same categories as the erect shrubs in the vigour studies. At the end of each growing season, beginning in 1986, thirty of these permanently marked specimens will be harvested and root crowns excavated to determine minimum age estimates.

Phytomass

A total of twenty-two 1x1 m quadrats were randomly located within ROW 1 and a further twenty-two in the control (Figure 4). In 1985 the total above-ground biomass was removed from these quadrats and sorted into species and their components. In 1986, eleven quadrats were denuded in ROW 1 and in the control (Figure 4). These phytomass samples were for all vascular plants in the shrub layer and below but no tree canopy material was included.

Tree layer phytomass was determined by sampling in 10 x 10 m plots. All *Picea mariana* and *Larix laricina* above 50 cm height were cut and subdivided into live and dead material within three trunk segments. In 1986 all dead standing trunks were also included. All material was weighed immediately and subsamples (10%) returned to the lab for drying and weighing. Discs were removed at the base of each tree and above and below any dead leaders along the trunk. Discs were returned to the lab for analysis. All cones were counted and the year of their production noted. This was accomplished by counting branch nodes back from the apical leader.

Propagules

Plant propagules can originate on-site as

part of the bank in the soil (i.e., seeds, roots, rhizomes, fragments) or as newly produced propagules generated from the remaining plants (i.e., seeds, rhizomes, spores, roots). Alternatively propagules may originate from off-site sources (i.e., seeds, spores). Both sources are important to revegetation of disturbed terrain.

On-site propagules were assessed by the removal of cylindrical (10 cm diameter) cores of 10 cm depth and subsequently maintaining them in a moistened condition in a greenhouse while monitoring for sprouting. Samples were removed from the mineral trench at 5 m intervals (Figure 4).

Off-site propagules were analyzed by two methods. First a set of snow sampling transects were established. At 10 m intervals a snow core was taken and when melted the seeds were filtered out. There are three transects extending for approximately 300 m each. Three surveys are done annually — early November, late February and mid-May. Seeds were also collected in the sixty litter-fall traps (see previous section on litter-fall).

Herbivory

A set of 140 live traps were laid out in fourteen lines running normal to ROW 3 (Figure 5). These lines were run twice in 1985 and three times in 1986. In 1986 the traplines were doubled in length and number of traps to include ROW 2. The lines were installed prior to clearing and then monitored during and immediately following the clearing operations of 1985 and 1986. All animals were weighed and marked for recapturing. Distribution and species composition data were collected during these studies.

Three trapping sessions were conducted in 1986. Session 1 (May) occurred before, during and after the clearing of ROW 2 and was designed to determine the immediate impact of this disturbance. Session 2 (July) and 3 (August), served to monitor the population throughout the summer season. The trench in ROW 3 was also trapped in July and August.

In conjunction with the distributional and compositional data, exclosures to prevent foraging activities were installed to determine the impact of small mammal use of vegetation in the trench area. A set of exclosures will be removed each year and biomass clipping conducted to determine differences between excluded and open foraging areas.

Conclusions

In a response to a lack of controlled experimental studies to evaluate reclamation treatments in subarctic environments, a long-term manipulative study was established. This study has been outlined in an effort to provide an example of the type of ecosystematic research that could contribute to the establishment of guidelines for reclamation programs associated with future northern developments.

Acknowledgements

The main financial support for the SEEDS project has been provided by Northern Oil and Gas Action Program (NOGAP). Support to myself and many of the students involved with the project has come from the Boreal Institute For Northern Studies and its Grant-In-Aid program. Additional funding has been provided to undergraduate students from the Royal Canadian Geographical Society. With a study of this nature there are, necessarily, a number of individuals and organizations that contribute in many ways to the successful operation of the project. I wish to thank them all for their dedication and perseverence. Several of these people are also conducting graduate research projects as part of the overall study. They include: Kevin Evans, Bonnie Gallinger,

Figure 5. Approximate locations of small mammal live-trapping network; SEEDS, Fort Norman, NWT — 1985 and 1986

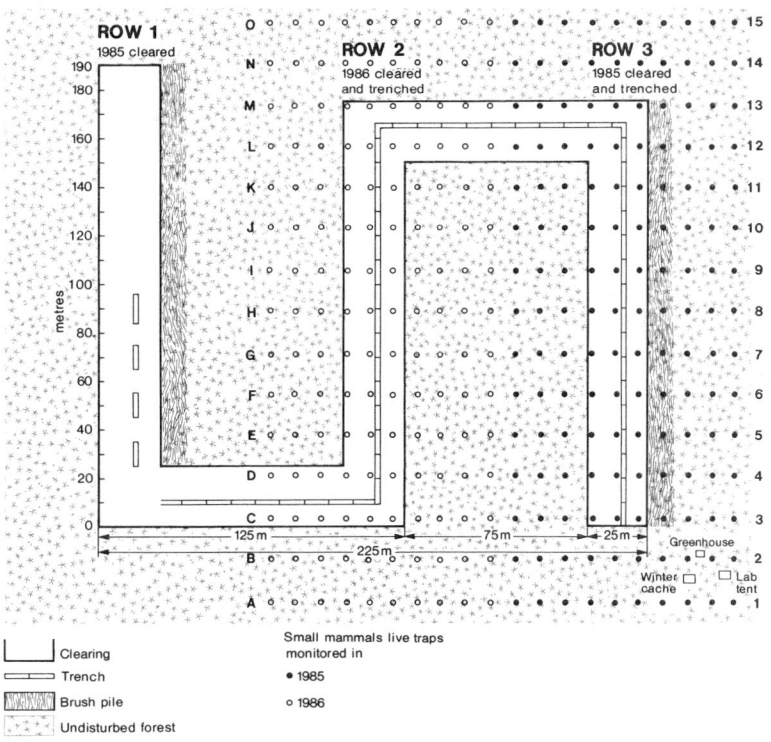

esa deGrosbois, Lynn Maslen and Ivan Shukster. At the beginning of the project Mat Fairbarns generated many ideas that have subsequently been persued and Steve Zoltai continues to make valuable suggestions and provide advice on many topics. Linda Kershaw has been involved throughout the life of the project and has made many contributions at all stages and is actively involved with field and office work. Don Wishart, Archie Pick, Tom Calnan and Leo Riendaeau with Interprovincial Pipe Lines Ltd. have supported the project by providing much appreciated logistics and advice that has been invaluable. Members of the Fort Norman Hamlet and Fort Norman Band Council have patiently reviewed requests for future work and provided much appreciated advice.

References

Bliss, L.C., Everette, K., Webber, P.J., VanCleve, R. and Viereck, L. 1981. Carbon balance in northern ecosystems and the potential effect of carbon dioxide induced climatic change. In: Miller, P.C. (ed.) Carbon Dioxide Effects, Research and Assessment Program, No. 015:4-16.

Canadian Soil Survey Committee. 1978. The Canadian System of Soil Classification — Research Branch, Canada Department of Agriculture, Publication No. 1646, Ottawa.

Evans, K.E., Kershaw, G.P., and Gallinger, B.J. 1988. Physical and chemical characteristics of the active layer and near-surface permafrost in a disturbed *picea mariana* stand, Fort Norman, N.W.T., Canada. — Fifth International Conference on Permafrost, Trondheim. (In press)

Higuchi, K., Etkin, D. and MacDonald, W. 1986. Trends in the variability of surface air temperature, Canadian Arctic. — Phys. Geogr., 7(4): 306-319.

Higuchi, K. and Etkin, D. 1987. Trends in year-to-year fluctuations in the surface air temperature, Canadian Arctic. — Atmospheric Environment Service, Downsview.

Kershaw, G.P. 1986. Studies of the environmental effects of disturbances in the Subarctic. — Norman Wells Project Monitoring Group Program, Third Annual Summary Report. Boreal Ecology Services Ltd., Yellowknife, pp. 115-125.

— and Evans, K.E. 1986. Soil and near-surface permafrost characteristics in a decadent black spruce stand near Fort Norman, N.W.T. — B.C. Geogr. Ser. 44: 151-166.

Malone, T.F. 1986. Environment 28(6)

Payette, S. 1983. The forest tundra and present tree-lines of the Northern Quebec-Labrador Peninsula. In: Morisset, P. and Payette, S. (eds.) Tree-Line Ecology, Proceedings of the Northern Quebec Tree-Line Conference. Collection Nordicana No. 47: 3-23.

Ross, M.S. and LaRoi, G.H. 1984. Structural dynamics of boreal forest ecosystems on three habitat types in the Hondo-Lesser Slave Lake Area of North Central Alberta in 1983. — Contribution No. 8, SEADYN

Rowe, J.S. 1972. Forest Regions of Canada. — Canada Department of Northern Affairs and Natural Resources, Forestry Branch, Bulletin No. 123, Ottawa.

Royal Society of Canada. 1986. Global Change: The Canadian Opportunity. — The Royal Society of Canada, Ottawa.

Steer, P. and Woo, M. 1983. Measurement of slope runoff in a permafrost region. — Can. Geotech. J., 20: 361-365.

Tuhkanen, S. 1984. A circumboreal system of climatic-phytogeraphical regions. — Acta Bot. Fen., 127: 1-50.

Worthington, E.B. (ed.) 1975. The Evolution of IBP. — Cambridge Univ. Press, Cambridge.

Younkin, W. and Martens, H. 1985. Evaluation of Selected Reclamation Studies in Northern Canada. Hardy Associates (1977) Ltd., Calgary.

Zoltai, S.C., Tarnocai, C., Mills, G.F. and Veldhuis, H. In press. Wetlands of the Subarctic regions of Canada. Wetlands of Canada.

Zoltai, S.C. and Pettapiece, W.W. 1973. Terrain, vegetation and permafrost relationships in the northern part of the Mackenzie Valley and northern Yukon, Canada. — Environmental Social Program, Northern Pipelines, Task Force on Northern Oil Development. Report No. 73-4.

NEW BOOKS FROM THE BOREAL INSTITUTE FOR NORTHERN STUDIES

ORDER FORM

Qty

Research and Monitoring in Circumpolar Biosphere Reserves *Norman Simmons, Milton Freeman, and Julian Inglis, Editors* ISBN 0-919058-65-5, ISSN 0068-0303, Occasional Publication No. 20 softcover, 8 1/2"x11", 75 pages A joint publication of UNESCO-MAB and the Boreal Institute for Northern Studies	$15.00
Knowing the North: Reflections on Tradition, Technology and Science *William C. Wonders, Editor* ISBN 0-919058-66-3, ISSN 0068-0303, Occasional Publication No. 21 softcover, 8 1/2"x11", approximately 150 pages	$24.00
Northern Lakes and Rivers *William C. Mackay, Editor* ISBN 0-919058-67-1, ISSN 0068-0303, Occasional Publication No. 22 softcover, 8 1/2"x11", approximately 120 pages	$25.00
Traditional Knowledge and Renewable Resource Management in Northern Regions *Milton Freeman and Ludwig Carbyn, Editors* ISBN 0-919058-68-X, ISSN 0068-0303, Occasional Publication No. 23 softcover, 8 1/2"x11", approximately 150 pages	$24.00
Northern Environmental Disturbances *Peter Kershaw, Editor* ISBN 0-919058-69-8, ISSN 0068-0303, Occasional Publication No. 24 softcover, 8 1/2"x11", approximately 70 pages	$15.00
Northern Communities: Prospects for Empowerment *Gurston Dacks and Ken Coates, Editors* ISBN 0-919058-70-1, ISSN 0068-0303, Occasional Publication No. 25 softcover, 8 1/2"x11", 100 pages	$15.00
Health Care Issues in the Canadian North *David E. Young, Editor* ISBN 0-919058-71-X, ISSN 0068-0303, Occasional Publication No. 26 softcover, 8 1/2"x11", 150 pages	$21.00
Kinship and the Drum Dance in a Northern Dene Community *Michael Asch* ISBN 0-919058-73-6 (hardcover), 0-919-058-74-4 (softcover), ISSN 0838-133X Circumpolar Research Series, Vol. I., Co-published with Academic Printing and Publishing 6"x9", approximately 140 pages	$27.95 hardcover $12.95 softcover
Small-Type Coastal Whaling in Japan: Report of an International Workshop ISBN 0-919058-75-2, ISSN 0068-0303, Occasional Publication No. 27 softcover, 8 1/2"x11", 120 pages	$20.00

Prices include postage and handling.
10% discount for ordering Occasional Publication Nos. 21 to 26 inclusive

Bill to: _____   Ship to: _____

_____   _____

Postal Code: _____   _____

Date: _____   Shipping will be book rate mail, unless otherwise requested:

P.O.# _____   _____

To order, send payment to: The Boreal Institute for Northern Studies, CW 401, Biological Sciences Building
The University of Alberta, Edmonton, Alberta, T6G 2E9
(403) 432-4999 or 432-4512

NEW BOOKS FROM THE BOREAL INSTITUTE FOR NORTHERN STUDIES

PREVIOUS PUBLICATIONS FROM THE BOREAL INSTITUTE FOR NORTHERN STUDIES

Qty

ORDER FORM

Qty	Title	Price
_____	**Snow cover as an integral factor of the environment and its importance in the ecology of mammals and birds.** Occasional Publication No. 1. *A.N. Formozov.* 1964. Translated from original Russian edition (1946) by William Prychodko and William O. Pruitt, Jr. 141 pp., illustrations.	$5.00
_____	**Catastrophic advance of the Steele Glacier, Yukon, Canada. A report on surveys conducted on the Steele Glacier from August 20 to August 23, 1967.** Occasional Publication No. 3. *L.A. Bayrock.* 1967. 35 pp., illustrations.	$1.50
_____	**On the edge of the shield: Fort Chipewyan and its hinterland.** Occasional Publication No. 7. *John W. Chalmers, Editor.* 1971. 60 pp., illustrations.	$2.00
_____	**An interdisciplinary investigation of Fort Enterprise, Northwest Territories, 1970.** Occasional Publication No. 9. *Timothy C. Losey, et al., Editors.* 1973. Results of a five-week archaeological investigation carried out as part of the NWT's 1970 Centennial Year celebrations.	$4.00
_____	**Among the Chiglit Eskimos.** Occasional Publication No. 10. *Emile Petitot.* 1981. Translated from "Les grands Esquimeaux" (1887) by E.O. Höhn. 202 pp., illustrations.	$9.00
_____	**Settlements of Northern Canada: a gazetteer and index.** Occasional Publication No. 11. *Roy Jackson Fletcher.* 1975. 136 pp., illustrations.	$4.50
_____	**The land of Peter Pond.** Occasional Publication No. 12. *John W. Chalmers, et al.* 1974. Series of non-technical articles describing the oilsand area of northeastern Alberta, its people and its history. 131 pp., illustrations.	$4.50
_____	**Climate of Arctic Canada in maps.** Occasional Publication No. 13. *Roy Jackson Fletcher and G. Stanley Young.* 1976. 48 pp.	$4.00
_____	**Consequences of economic change in circumpolar regions.** Occasional Publication No. 14. *Ludger Müller-Wille et al., Editors.* 1978. Papers of the symposium on unexpected consequences of economic change in circumpolar regions at the 34th annual meeting of the Society for Applied Anthropology, 1975. 269 pp., illustrations.	$10.00
_____	**Wildlife of the Mackenzie Delta region.** Occasional Publication No. 15. *A.M. Martell, D.M. Dickinson, and L.M. Casselman.* 1984. Synthesizes information from published accounts with that of more recent papers in an annotated list of all invertebrate wildlife species that occur or have been reported in the Delta region. Also includes vegetation. 214 pp., maps.	$15.00
_____	**Between two worlds: the report of the Northwest Territories perinatal and infant mortality and morbidity study.** Occasional Publication No. 16. *D.W. Spady et al.* 1982. The health of infants born in the NWT in 1973-74 is analyzed with respect to the socio-economic and cultural environment, nutrition and health care. 271 pp.	$15.00
_____	**A time for burning.** Occasional Publication No. 17. *Henry T. Lewis.* 1982. The use of fire by native peoples of northern Alberta to transform and maintain their natural environment is examined. 62 pp. Complements the film *Fires of Spring*.	$5.00
_____	**Odyssey Northwest: a trilogy of poems on the Northwest Passage.** Occasional Publication No. 18. *Gerald St. Maur.* 1984. These poems deal with the explorations of Martin Frobisher, Henry Hudson and Sir John Franklin. 123 pp.	$15.00 hardcover $10.00 softcover
_____	**Keeveeok awake: Mamngugsualuk and the rebirth of legend at Baker Lake.** Occasional Publication No. 19. 1986. Catalogue of 20 drawings, 5 prints and 1 wall hanging by Baker Lake artist Victoria Mamngugsualuk. Includes French translation insert. Jointly published by University Collections and the Boreal Institute.	$15.00

Prices for these books do not include postage and handling.